An Evolutionary Fairytale

EVOLUTION

A Fairytale for All Ages™

by Nick Hetcher

The Handbook That Destroys Evolution

Every biblical creationist should have this book to share with believers and non-believers alike

1ˢᵗ Edition

Cover Photo
(KellPics / Pixabay.com)

I0490454

CHARLIE (Charles Darwin) **WAS WRONG...** and the contents in this book will prove it without a shadow of a doubt. Evolution keeps multiple millions of people from an eternal relationship with our creator God, which should be seen as a much greater threat than all the world's diseases combined, yet people still embrace it. How sad. This book will change that viewpoint for many. The content in this book absolutely proves that God *does* indeed exist, and that evolution is nothing more than a fairy tale (or "*fairytale*" as used in this book). If you believe that God created the world, this book will help you share it with your believing "and" non-believing family and friends. If you still believe in evolution, get ready to be challenged. BIG TIME. This book is NOT about bashing evolutionists/atheists, rather it's about completely **destroying the "theories" of evolution** themselves, and exploring deeper into *Intelligent Design* and the God who created and loves **YOU.**

An Evolutionary Fairytale

DEDICATION: To my beautiful wife, Lynn Collette, who has supported, challenged, and helped me a lot in writing this book. I also want to thank my *incredible* family (mom, kids, grandkids, brothers, etc.) and friends for your love, encouragement and support, and my son Nicolas who also gave me assistance in writing this book. You guys are all AMAZING!! And, to all you cool creationists out there, may this book help you, your family, friends and co-workers... in life's journey. Most thanks goes to our Awesome Creator God!!

MUCH THANKS TO: Steve Hetcher, Jeff Peterson, Jeff and Kim Schwarz, Julie Musial, Rollie Nollenberg, Mark Hetland, and Mark and Theresa Westphal (the proofreading queen) who went above and beyond the call of duty in helping to edit and proofread this book. You guys are all awesome!

A word from the GREEN BAY PACKER'S pastor:

"Nick somehow finds a way to take a very complex issue that's been confusing and misleading people for decades, and make it completely practical and understandable. If you've had questions or confusion about evolution - this book is for you. 'Evolution: A Fairytale For All Ages' is a game changer."

- Shawn Hennessy (Senior Pastor at Life Church Green Bay)

Is Evolution Really Just a Fairytale?

Is it the real deal? Or is it one of the biggest shams in history?

Either *Evolution is a Fairytale,* **or** *the Existence of God is.* There are *no* other real choices. They are complete opposites.

You'll find that **the contents of this book will not only** *completely destroy* **the** *unproven Darwinian Evolution Fairytale,* **but also all of the** *"other"* **theories of evolution as well. Don't get me wrong, I love the evolutionist and atheist (the people), but I absolutely hate the horrendous fairytale** *theories* **that they believe are true.**

You see, ALL of the theories of evolution are *feeble fairytales,* and contrary to popular belief among evolutionists, they have *no* **actual** *evidence* **to prove their completely random chance theories.** Don't believe me? Well, let me *prove* it to you in this book.

Evolution is a wicked fallacy, yet sadly, millions of well-meaning people simply believe what they were told about it (with no actual supporting *evidence),* **in two weeks in a high school class.** Since Darwinian evolution claims life came into existence by *billions and billions* of *random chance processes* over millions of years, and *not* by a *supernatural creator, don't you think actual EVIDENCE is necessary to prove the grandiose evolutionary theories?*

THIS BOOK IS A <u>RESOURCE</u> BOOK, *NOT* A NOVEL. *Skipping* and *Skimming* is permitted. Even encouraged at times. :-)

<u>VERY IMPORTANT</u> REQUEST... I need YOUR help. After reading this book, *please make "sure"* to go to this book's page on <u>Amazon.com</u> and **write a short REVIEW.** This can help us reach *many* more people with the truth of creation and God's eternal plan. Also, please share this book and our social media posts with friends.

In the Beginning
GOD Created
Heaven and Earth

Genesis 1:1

The fool says in his heart,
"There is no God."

– Psalm 53:1

Table of Contents

EVOLUTION

A Fairytale for All Ages™

by Nick Hetcher

WHY THIS BOOK... This book was written primarily as an *apologetics (defending the faith)* **resource for God believing** *creationists,* **as well as those** *fence-riders* who are just not sure about their stance on the whole *creation vs evolution* debate. **It's for all adults of course - but maybe even more importantly, it's for** *pre-teens, teenagers, college aged youth, and millennials.* **Every pre-teen to adult needs to know this information.**

With the evolution fable being taught in schools everywhere, youth need to hear the *truth* about creation. The information is this book can help change their (your) evolutionary worldview. I hope parents will get it into their kid's hands and encourage them to read it. Even bribe them to read it. Hey, why not? It could change their eternal destiny. I hope it will stir their (your) emotions, and challenge their (your) brain, and get them (you) sharing the *real creation* story with *everybody.*

I know the Bible says God created the Earth in seven days. I'd also like to believe there may have been an eighth day where God created *Coffee, Bacon,* and *All You Can Eat Crab Leg Buffets.* ;-)

Joking aside, personally, I categorically abhor the *false teachings* of the erroneous *theory of evolution,* primarily because of what it has done to distract people from the *real truth* and *eternal life with our Creator*. Hundreds of millions of Christians around the world agree.

The evidence for *CREATION SCIENCE* is clear and straightforward. In a word, it IS *scientific*. Evolution on the other hand is *not* only *NOT* Science, it's "*ANTI-SCIENCE*" as you'll learn in this book. Yes, you read that right.

Charles Darwin's biggest disciple and bulldog, **Thomas Huxley,** *admitted that creationism was in no way illogical. When an explanation is logical and has the evidence in its favor, and the rival theory is illogical and lacks any substantive evidence, then the conclusion is obvious.*

"*'Creation'* in the ordinary sense of the word, is perfectly conceivable. I find *no* difficulty in conceiving that, at some former period, this universe was *not* in existence; and that it made its appearance in six days... in consequence of the volition of some pre-existing Being." - *Leonard Huxley (Evolutionist. Life and Letters of Thomas Henry Huxley, Vol. 11 (1903), p. 429)*

Be prepared to learn a lot of information about evolution that was *intentionally hidden* from you. And yes, many times out of sheer ignorance, and lack of *real* science. I did my best to be as accurate as possible, if I misquoted anybody, it was *not* intentional.

I only touch the surface on the topics of evolution in this book, yet still enough to give you the information needed to support *intelligent creation* and *crush* the theory of atheistic evolution. I encourage you to dig deeper and find much more information on these subjects.

The deeper intention of this book is to expose the utterly ridiculous *theory of evolution fairytales* and to lead people to the knowledge of the *truth of creation* (much of it is true science found throughout the

Bible, believe it or not). That statement really irritates skeptics and is ridiculed by evolutionists and atheists alike, however, *they are incorrect and I'll prove it to you many times over in this book.*

The content in this book is *not* to attack most atheists, agnostics, and/or evolutionists themselves (OK, maybe those who spearhead the charge, a wee bit).   It's an easy to read book that challenges your intellect, common sense, and evolutionary worldview *(if you're an evolutionist)* - and hopefully leads you to studying, understanding and applying the wisdom of *the greatest book of all time, the Bible.*

Warning: If you are an atheist, agnostic and/or devout evolutionist (and maybe even a few creationists)… the content in this book and my slightly sarcastic writing style may offend you a bit. But then, I'll bet many books or articles defending creationism may offend you somewhat. If you're *not* a creationist and you're not *open minded,* you might as well put this book down now. Seriously.

SO, HERE WE GO…

STOP DRINKING THE "EVOLUTION KOOL-AID"… The different ***Theories of Evolution*** (Cosmic, Organic, Macro, etc) are some of the ***Biggest Boldface LIES Ever Told to Mankind.*** Please allow me to elaborate.

Most people who do believe the obtuse and *non-scientific theory of evolution* lies, really have *no idea WHY they believe them,* other than they were told in a science class that evolution is true (yes, even most scientists as you will learn *don't* really know *why* and *how* so-called evolution is allegedly true, and have *no* actual evidence at all, even though many claim they do). If you believe in evolution now, or are just not sure if it's true or not - this book is going to teach you how to ***STOP DRINKING THE EVOLUTION KOOL-AID.***

This book will expose and debunk many of the evolution fairytales, and teach the *truth* about *God's* creation. And, to be clear right up front, **I don't have a college degree in any of the sciences. I'm *not* a scientist. I don't even play one on TV. But then, neither did Charles *(Uncle Charlie)* Darwin have a science degree or play a scientist on TV.**

As I began writing this book, I knew there would be vehement opposition, starting with its deliberately provocative title. Still, I felt a very strong passion and duty to complete it and get it out to as many people around the *(created)* world as possible. (Jab, jab, uppercut. Evolution down for the count). It will encourage and strengthen creationists in their beliefs, as well as reach many who don't know their Creator... *yet.* I hope you will help me reach these people, too. You creationists will help me, right? Right? :-)

You see, **the *Theory of Evolution* can actually be *ANNIHILATED* very easily** as I'll touch on in the next few minutes here, and then we'll get into it much deeper so there's no doubt many of you will agree that **it's *a fairytale for all ages,* and *must* be exposed as such.**

Let's start with a brief overview of why the alleged *Darwinian theory of evolution* (or *"Monkey Science"* as I like to call it), is flawed more than a corrupt politician.

For starters, the **scientific *impossibility* of the incredibly complex "first cell" ever forming** and actually staying *alive, evolving* and then *reproducing* - is a huge evolutionary theory killer. Another **best *proof against evolution,* that *can't* be disputed, and literally *crushes* the apathetic theory... is the *lack* of any real *evidence* for...** *"transitional* **fossils."** This may be **the best evolution slayer of all.**

You know, those fossils (in-between) *different* kinds of animals and plants; such as *in-between* a fish and a land animal, bird and dinosaur, monkey and man, etc. Or apple tree and a rose bush, etc.

Sure, a handful of *fakes* have been presented and *falsified* over the years, but *no* real *transitional fossils* have ever been found. Think with me for a moment - **There are estimated to be in the *trillions* of *fully-formed* fossils in the worldwide fossil record** (for the record, a high percentage are small sea creatures) **so there absolutely MUST be many *trillions*, for evolution to be true.** Yet, we find NONE.

It's utterly *impossible* for them to disappear, or hide somewhere to be discovered in the *future*, as some like to say. IMPOSSIBLE!! That is *not* a credible, true or sane answer. Could the complete *lack of evidence* in the *fossil record* be any clearer that evolution is only a fable? **Even a 5 year old can *destroy* the evolutionary theory with this one monumental evolution buster.** Actually, this book could end right here and right now.

*IN OTHER WORDS... **Molecules to Man Evolution* NEVER happened, does *not* happen now, and will *never* happen in the future.** The bogus *theory of evolution* shell-game is no more than cheesy *Science Fiction* and is *COMPLETELY IMPOSSIBLE*.

Leading evolutionists and science teachers won't accept this and say you should believe their theory anyway, and you should ***bow to their self proclaimed superior intelligence.*** *"Do not question, do not doubt, do not think. Just allow the brainwashing to take control of your mind."* Now, just because some scientists chase their tail around in circles like a monkey, doesn't mean you should.

Starting to see my extreme dislike for the full-of-holes theory? But Wait, There's More. Most people who simply believe in evolution don't even know this fact about transitional fossils because it's been *hidden* from them. If that's you, now you *do* know so there's *no more excuses to defend evolution.* FYI, most science teachers don't even know this.

However, much of what the egotistical leading evolution gurus state as fact about evolution, is in reality, complete *intellectual nonsense.* Many of them are not all as bright as they'd like you to believe, when it comes to evolution and intelligent design creation. All you need to do is ask them to provide solid *evidence* for their outlandish statements, and you'll quickly uncover their *nominal understanding* of real creation science and evolution. A big mistake would be to simply take their words at face value. *Challenge them.*

ARE YOU KIDDING ME? When I hear them talk about things for example, like a *bear or cow evolving into a whale,* I just have to laugh. Hard. (*Yeah, the bear jumped into the ocean one day and swam and swam and swam for a million years until it finally became a whale. Oh, and where are the transitional fossils to prove it?*). I refuse to even talk about it in the *animal section* of this book because it is so ludicrous. **I for one AM going to *question* their theories. So should you. I AM going to *doubt* them all until I see actual *evidence.* So should you. And, I AM going to *think* for myself. And, so should YOU.**

Personally, my quest for the truth of creation lead me to the *certainty* that *evolution is not true* and *creation is true.*

The ONLY other possibility other than evolution is *Intelligent Design* by a "*Supernatural*" being. Most just call Him... GOD!! End of Story!! Or is it? Not satisfied? Let's continue...

Just these two topics "alone" (the *first cell* forming by complete random luck, and *no transitional fossils* ever found), *completely* *destroys any* **and** *every* **theory of evolution;** be it *Darwinism, Neo-Darwinism,* the newly branded version called *the third-way (same lame stuff just packaged differently),* and a few other oddball theories.

Simply *"believing"* **or** *"thinking"* **something...** such as the universe evolved originally from *nothing,* or that we all exist by some *strange fate, random chance,* or *pure luck...* **does** *not* **make it true.** Right? That's foolish, naive thinking. Only *solid, indisputable scientific evidence* could possibly make it true.

On another note... here's a Bible verse worth knowing... **"The fool says in his heart, *'There is no God.'"*** - *Psalm 14:1* (It may be offensive to some, however, what if it's divine insight from our creator to get our attention? What if it was said in love? Actually it was. :-)

Believing we are all here by *ACCIDENT* - meaning that the first cell somehow magically evolved *(crawled out of a slimy mud pit),* and that over *really* long periods of time this single cell eventually evolved into a *fully-formed* heart, brain, lungs, muscles, skin, eyes, mouth, central nervous system, etc. - **is only science** *fiction.*

MINI SCIENCE LESSON... **Unless ALL of these** *essential* **body parts, and so many others, are** *in their PROPER PLACE in the body,* **and** *FULLY-FORMED* **from the** *very first few minutes* **after birth... LIFE COULD** **NOT** **EXIST for a million different reasons. Don't be told otherwise as that's a complete lie.**

Then there's the *massive problem* **of a mate of the** *opposite sex* **appearing at the** *same time* **and in the** *same geographical location,* so they could marry and have a clan of kids like the *Duggars* (You know, that TV show with the 19 kids). Just these few facts alone absolutely *destroy the evolution hypothesis,* but let's not stop here while we're having so much fun.

Or, that the *first* species, over long, long periods of time, evolved into multiple thousands and more of completely *different* species. That first single cell eventually evolving into *whatever:* an acorn growing into a rose bush, a fish evolving into a land animal, or a monkey into a human... are only *fairytales.* Lies, lies and more *uneducated, unadulterated lies and scientific impossibilities.*

TALK ABOUT CRAZY... Actually, **this crazy evolutionism theory goes as far to say that even *HUMANS* are RELATED to all *PLANTS,* too.** Did you know this? Ludicrous, right? Yes, apparently the oak tree with the broken swing in your backyard is a distant cousin to all of us. No joke, that's exactly what the looney *theory* claims, however, **any *authentic* evidence to support this ridiculous theory is completely lacking.** Ain't none folks!! Again, there *HAS* to be thousands of human *transitional* fossils actually *discovered in existence,* not some *pencil drawings* or made up stories about them.

And, in many cases for sea creatures and plants - millions to multiple billions or more of transitional fossil records *must* exist. Sadly, for the evolutionists, there are *none.*

EVIDENCE... **is what you want, right?** Well, that's what this book is about. We'll *prove* that the spurious theory of evolution is unconditionally *false* - through and through to the bitter end.

Let's kick evolution around a little more before we really destroy it for good... This is fun, hey? Well, at least for creationists it is.

"Variation" **within the *same* species *does* of course occur** (*i.e. a bird beak changing slightly, different kinds of dogs, cats, insects, fish, birds, etc*). **This is NOT evolution,** but rather is merely known as *"Adaptation."* Meaning, animals *adapt* to their surroundings, changes

happen, however, always keep in mind... **a dog *always* remains a DOG.** Nothing else. EVER.

Many evolutionists like to call this *"**micro**-evolution,"* but **it is *not* evolution at all,** that's just what *they* named it. **It does *not* prove evolution from** *one kind to another kind* **in the past *or* present in any conceivable way** (This they call *"macro-evolution."* Such as a cat becoming a dog, dinosaur to bird, ape to human, etc).

No matter how much some evolutionary scientists want to believe it, or want *you* to believe it, *adaptation* **is certainly *not* a precursor for** *macro* **evolution,** as *lack* of any credible *evidence* plainly reveals.

Again, **I mention *evidence*, because that's all evolutionists need to provide to prove their theories. Without *evidence*, and a whole lot of it, their so-called theories are absolutely *worthless*.**

Either the universe and all living creatures and plants evolved naturally by *unintelligent, random chance, atheistic evolutionary natural processes* **- OR - they were created by a** *Supernatural, Intelligent Creator* **(God). It's *not* both. You only have these *two* choices.** *Aliens* **seeding planet Earth with life is *not* an option as some actually and unbelievably claim it to be.**

THE PROMISE of this book... To expose the erroneous *Theories of Evolution*, thus validating the *only* other creation option, *intelligent design* by an intelligent, supernatural, loving creator, most call God.

Creationists **will want to read this book** - to learn more about why you can believe you were created for a *purpose,* and to share this amazing information with your creationist family and friends who should know it better, as well as all of your skeptic family and friends. *Evolutionists* and *Atheists* will want to read this book so you can finally learn that what you were taught, the contrived theory of evolution - is a complete and mindless *fairytale*. A silly fable of *non-science brain-washing*. If you think it's any more, I've got a bridge for sale... At *least* read it to **compare the two options more equally,** and not only know about the jaded anti-God evolutionism worldview.

See, even Uncle Charlie had *serious doubts. Apparently, a "cold shudder" ran through him. Now that's something I certainly wouldn't want happening to me. Would you?* **"Oh no, not a COLD SHUDDER. Everybody run for the hills, Uncle Charlie just got a 'Cold Shudder.' "**

"Often a *cold shudder* has run through me, and I have asked myself whether I may have not devoted myself to a *fantasy."* - *Charles Darwin (Life and Letters)*

Think about that statement. *Cold Shudders* and all, it says a lot about his doubts, the doubts that multiple millions of people hang their lives on. **If Darwin doubted the theory of evolution, shouldn't we?**

Alrighty then, let's get into this... It's funny how when made up stories are written for children, we think of fables or fairy tales, but when they're written for adults, they're called, *"the theory of evolution."*

Coming up next, some wild and crazy stories written by Uncle Charlie and his evolution science friends. **You'll learn how evolution is not only *not* science, but rather *anti-science* on so many levels. Please keep in mind that this book exposes the fallacies of all the *different* kinds of evolution, starting with *Cosmic Evolution,* it's *not* all about Charlie. Sorry Charlie. ;-)**

Stick with me as we dig even deeper to uncover one of the *Greatest Hoaxes of All Time... Evolution.*

*O*nce Upon a Time... *NOTHING* decided that he wanted to become *SOMETHING.*

Not just *Anything* but *Something* really HUGE, like - the *Universe* and *Everything* in it. I guess you could say *Nothing* became the first *Mr. Universe...* And, you thought it was Arnold Schwarzenegger. Eventually, as the story goes, after about 10 billion years, *Nothing* supposedly became the *first living cell,* and then the *first human* formed after another almost 4 billion years. *Nothing* had really, really BIG dreams you see.

But first he needed to find *Mrs Nothing* and start a *Nothing* family...

So, according to the *misleading opinion of evolution,* apparently *dreaming, thinking, emotions* and even *love* and *starting a family* - started from absolutely *nothing,* too. Oh yeah, I almost forgot, somehow *Mrs. Nothing* had to magically start from thin air, at the *same time* and in the *same location* as *Mr. Nothing,* as well.

Wait a cotton pickin' minute... Are you saying that the entire universe and planet Earth started billions of years ago from absolutely *nothing?* Give me a break, bruh. (Yeah, just trying to be hip using a fleeky old saying with today's way to say *bro.* And, now cool or groovy using *fleeky* instead. I better give up now, who am I kidding? I'm a *dork,* or worse yet, a *cornball,* a *krod,* or a *flapdrol.* Hey, stop judging me just 'cause you think I'm a duck. Haha, lol and all that stuff. Thank you *Urban dictionary* for just wasting 15 minutes of my time instead of writing this important book. All that to say... whatever age you are, don't let a few old or new sayings I sprinkle throughout this book bother you, get over it and just learn from the overall content about creation and evolution. Remember, this is a *resource* book.

Seriously, even *every* living creature today, originally started millions of years ago from absolutely *nothing, too?*

But, but... **how in the world can** *something* **like the entire universe and all living things in it, start from** *NOTHING?* That doesn't make any scientific sense whatsoever. Am I right or am I right? Deep down, most of you know this is right.

And, **if you think that there was** *always* **something there, then where did** *that* **something come from, or** *how* **was it always there?** As we'll discuss later, the *mathematical probabilities* of this happening *naturally* by *chance* are far beyond zero. OK, just zero. :-)

When you think deeply, you should agree that "*Supernatural***" Creation takes a** *lot less Blind Faith* **than believing in "***Natural Proven-Less Evolution.***"**

Francesco Redi *disproved* **the fraudulent theory of** *spontaneous generation* (life evolving from non-life) in the 1600s. And, **the famous scientist (and milkman)** *Louis Pasteur* **also proved that** *life cannot possibly come from non-life,* **but** *can ONLY arise from life* (**i.e.** *Biogenesis*).

I find it hilarious how evolutionists love to describe what *GOD isn't,* **and what** *NOTHING is.* They can't *see* God so they say he *doesn't* exist. *They can't see evidence of evolution between species either* (such as dinosaur to bird), yet claim it *does* exist. Huh?

Most Evolutionists believe... the universe started from absolutely *nothing (Cosmic Evolution), that* somehow *exploded.* How *nothing* explodes nobody has a clue since it's, well... NOTHING. And, it exploded, how? *HELLO, anybody in there?* Yes, people actually buy into this crazy *no-proof* theory (fairytale), but won't allow themselves to believe in an *intelligent designer.* See why I wrote this book?

Then, billions of years later, life on Earth was magically created somehow by **random chance chemical reactions,** and continued to the present day by gazillions of ever-so-strange acts of *random mutations.* (FYI... Mutations do *not* "add" anything good as would be *required* for evolution to be true. **Mutations are almost always destructive.** More on that later).

WHAT? Could **Charles Darwin and the Evolutionists** (Gee, sounds like a 70's punk rock band) be wrong about the evolution theory? In a word... ABSOLUTELY YES. OK, that's two words. Ya got me on that one.

"Scientists who go about teaching that evolution is a fact of life are Great Con-Men, and the story they are telling may be the *Greatest HOAX ever!* In explaining evolution we do not have one iota of fact." - *Dr. Newton Tahmisian (Atomic Energy Commission)*

To Most *Thinking* People (the key word here is, *thinking*)... *HOW and WHY we exist may be among the most important questions in life.* The juxtaposition of *intelligent creation* and *random chance evolution* (or *evil-ution* as many Europeans say) is what much of this book is about, with **the main focus on the *truth* of creation, and what that actually means to us. What I'm saying is that there is a** *purpose* **behind it all. Evolution** *doesn't* **offer a purpose. God's** *Intelligent* **Design does.**

You'll learn how the distorted theory of evolution has *blinded* millions upon millions of good people to the fact that they were created by an all powerful Creator. **Creationists can't stand by any longer, we *must* share the real truth of creation that needs to be heard and shared to *everyone*, everywhere. It's time to step up our game!**

IN THIS BOOK WE'LL EXPLORE HOW... *Modern Science, Irrefutable Historical and Empirical Evidence, Mathematical Probabilities* and *Common Sense* all *prove* without a shadow of a doubt that **the universe and every living *creature* and *plant*** on Earth **could ONLY have been created by an *intelligent* designer,** and could *not* possibly *(in any sense of the imagination or cleverly invented stories)* have evolved by *un-intelligent random chance evolution* over millions and billions of years.

Hopefully you're open-minded and ready to be challenged to learn *real* creation science. If you are, and are currently on the *evolution team,* get ready to have your mind blown and to potentially change sides to join **TEAM CREATION.** If you don't, it's certainly *not due to lack of evidence disproving evolution over and over.* You are just buying into *unproven* evolutionary theories as if they were somehow proven. That's a big mistake. Many evolutionists act and talk like they have all the evidence. Don't be fooled, they don't.

The way we have all been taught from childhood (our *worldview*) holds many undecided people and most ardent atheists back from accepting creationism. And, maybe the overwhelming *lack* of *viable evidence* for evolution doesn't mean anything to you because you were taught in school that, **Given Enough TIME,** evolution will reign over creation (This book will prove this shaky *"given enough time"* idea to be utter nonsense). In reality, **TIME is the enemy of evolution, not it's cohort.** More on this in a bit.

Maybe **you** *choose* **not to believe** the truth about creation for *another* reason. Maybe **being accountable to a God is** *not* **what you want,** but rather being in control of your own destiny. Don't buy into that deception. **There's incredible** *freedom* **in God,** many of you just don't know it… yet. The good news is, you can.

IF GOD IS REAL - wouldn't you agree that this is *the most important* **thing you should know about in this life?** I hope you'll give me the opportunity to *prove* that the suspect theory of evolution is *not* true, and that there *is* a God who created us and loves us.

IMPORTANT NOTE: So that you don't think I *stutter* when I write, some things will be *repeated* (repeated) in this book (OK, my humor lacks at times. Agreed), and the ***formatting*** *is also different* than in most books (i.e. you'll see words *italicized,* in **bold,** BIG HEADLINES, weird spacing, and underlined) - this is mostly *intentional* because I want each topic to stand on its own. Many *skimmers* and *skippers* will only read some parts and skip others so hopefully this formatting will make it easier to see key topics and points. On another note, **you may get tired of me bashing the** *dishonest theory of evolution* **again and again (and** *again***) - this too is for those who jump around between topics, and to drive the point home.** From personal experience, most atheists don't have a problem trashing creationism and creationists at every turn. So, to be fair, they should be tolerant and not be offended if we trash their almighty *random chance theories*. One other thing - **when you see the word** *evolution,* **we may be talking about just "one"** *(or possibly more than one)* **of the (different)** *theories* **of evolution.**

THE THEORY… (or maybe better stated as the *"ASSUMPTION"*) **of Evolution is…** as you're about to learn - **a collection of cleverly packaged** *half-truths* **and downright** *lies* **that have nothing** *real* **to back them up.**

If you listen close to the *evolution fairytale,* all you hear is a bunch of *surface level jibber-jabber.* Even though a lot of ideas and theories are tossed around like a fastball at a Yankee's game, there's no real substance when examined closer. So, I encourage you to open your eyes, ears and heart as if your *eternity* depends on it. Actually, it does.

Evolution truly is a *Fairytale for All Ages!!* But *much worse,* even innocently **believing these flagrant lies and half-truths can have** *eternal consequences* **for you and your loved ones.**

Let the flaming evolution extremists laugh at this book (and at creationists) all they want. They most certainly will. **They are *wrong* about the evolutionary story of creation, and we're gonna put all the antiquated, idiotic theories of evolution to the test in this simple, easy to read and understand book.**

Yeah, I'm talking a lot of smack. Don't worry, I'll back it up. In fact, **by the end of this book you'll know *much* more about the *real* science of creation than any evolutionary scientist or college professor** would like you to think they know. I'm sure some tempers are flaring about now, however, let truth prevail. Isn't *truth* what we *all* really want anyway? Unsubstantiated talk is cheap chatter.

Dear Evolutionist, Skeptic, Agnostic and Atheist...

You may be ready to stop reading this book already or at least send me a nasty email. Some of you are pretty angry with all the trash talk about evolutionism so far. Well, fear not, there's a lot more coming, and if you stick around - *I GUARANTEE to scientifically and mathematically disprove all of the theories of evolution.*

PLEASE UNDERSTAND - *this book is NOT about bashing you, or anybody* for that matter. OK, maybe some teachers, professors, and prominent outspoken evolutionists. And, of course their notorious commander and chief Mr. *Charles Darwin.* Darwin is long dead anyway so I'm not going to hurt his feelings. The vast majority of creationists have nothing against the people who believe in evolution, in fact we rather really like (and yes, even love) most who don't agree with us on this important issue, even though it may not come across that way at times when discussing evolution and God related issues. So, please keep this in mind as you read this book or the next time you're discussing evolution and intelligent design with a creationist.

The first goal of this book is to *expose* **the erroneous, whimsical** *theory of evolution,* **thus proving** *supernatural intelligent purpose creation* **to be the absolute ** *truth* **about Creation.** Then, we'll discuss why the Bible can be *trusted,* and why Jesus is the *only* Savior of the World. (Yes, a very narrow minded comment, *if it's not true.* I get that. But *not* so narrow minded if He is who He says He is - that being God in the flesh and the Son of God).

Well, you gotta give the *natural evolutionary scientists* credit for trying and losing again and again and again. But seriously, **they're** *never* **going to find the answers in** *nature,* **but rather in the one who** *created* **nature.** Yes, God is *supernatural* and yes He's real. And no, He's not a woman. Who cares? It doesn't mean human women are any less than men, they are equal and *equally loved by God.* So are *all* nationalities. We *all* bleed red and are *all* related to *Adam and Eve* and are *not* related to a coconut or a banana or even a cute little chimp. **Recent DNA studies** *prove* **we all started from one man and one woman, several** *thousand* **years ago,** *not* **millions.**

For the Record... **I'm** *not* **looking for your thoughts and arguments** as to why evolution is right and intelligent design is wrong. I'm sure I've heard most of them by now, and frankly even though I'd like to debate individually, sometimes, I just don't have the time and energy. You see, in most cases I have experienced that *most atheists and evolutionists won't really listen* to me or any creationist anyway. (Please see my **"How to Beat ANY Atheist/Evolutionist at the...** *Creation vs Evolution* **Game,"** towards the end of this book). Many other creationists will be more than happy to debate you, but not me.

Please try to hold back judgement on me and other creationists until you have read this *entire* **book.** Hard to believe now, but many of you will change your mind before finishing this book and become Creationists, and much more importantly, even *Jesus* followers *(the ultimate goal of this book).* We'd love to hear about your conversion. Please post it on our FB page: *facebook.com/CharlieWasWrong*.

The Original "Fake News"

The **EVOLUTION** *"ASSUMPTION"...* just may be the genesis of *Fake News.* ;-)

Don't be fooled by a few overconfident evolutionism believing zealots in high places, and their babble talk about deranged evolutionary theories - but rather *challenge* **them.** Be *kind and respectful,* but absolutely do challenge them. Challenging questions are fair game and can be found later this book.

The irresponsible theory of evolution should *not* **be allowed to get away with its blatant lies about** *microbes to man, particles to people, goo to you, cell to selfie,* **and** *chimps to chumps evolution.*

The evolution religion should *not* be taught in our schools, if the truth, *supernatural* creation (by God) is banned. Why do we taxpayers put up with this? Be bold and ask passionate evolutionists to try to provide *real* evidence to the *lies* being taught in schools. Don't worry, though they may try really hard, they *cannot* provide *any real evidence.* It's all a big *house-of-cards.*

Fact is - the high school and college science textbooks are 30 to 50, even 100+ years *behind* the times on how science has *disproved* evolution in so many ways. Many evolutionary scientists know this and are not saying anything. Sadly, many teachers are still buying into and teaching this outdated trash. Many are unaware of its falsehoods.

Are we living in the *Dark Ages?* Seriously, evolution is a mammoth joke on the public at large. Well, billions of creationists worldwide are *not* laughing and… **now more than ever is the time to *speak out against evolution.* We can no long rest on our laurels!!**

Evolution (aka: *transmutation*) **is the biggest and most cruel *joke*** the world has ever heard. And to think so many unquestioning people take it seriously and just believe whatever they hear or read without questioning it.

STOP being a Sucker. Stand up for the TRUTH.

Creationists can't sit by quietly and allow the people we love be deceived by it any longer. There really is an *Intelligent Creator* of the universe. And, as weird as it may sound to some of you, he loves YOU, and in this book we're going to *prove* that evolution is *not* how we all came to exist, but nothing more than a grand scale parlor trick. Rather, **we were created for a *purpose* and *incredible eternal destination,*** by a real, supernatural, all powerful, all knowing and loving CREATOR. Know this... **YOU ARE "NOT" AN ACCIDENT OF NATURE!! YOU ARE PRICELESS!!**

"Evolutionism is a fairy tale for grown-ups. This theory has helped nothing in the progress of science. It is *USELESS.*"

- *Bounoure, Le Monde et la Vie (Director of Research at the National Center of Scientific Research in France - October 1983)*

If parts of this book make you angry, please don't take it out on the messenger. Don't take it *personally* either. Rather, why not point that irritation towards the evolution scientists (and text books) who teach and perpetuate this outrageous theory as if it's the truth, and **challenge them to** *provide hard evidence* **for what they teach.**

Evolution only happens in the *imagination,* **never in reality.** The exciting information in this book will *prove* beyond any doubt that EVOLUTION between different *kinds,* is *no more than a fantasy* as even Mr. Darwin himself eluded to.

In the Beginning

"In the beginning, God created the Heavens and the Earth" (Gen. 1:1)... After that, everything else was - *Made in China.*

That's obviously a joke, I'm *not* a racist. (It's too bad I actually have to write a disclaimer because of all the super sensitive, intolerant people now days. Ruins a good joke).

Many proponents of the preposterous *"theory,"* will not like this book *(or that joke)* because it exposes a myriad of lies, fallacies and *non-*scientific facts that the lackadaisical theory of evolution foolishly cries out, in which so many people are blindly convinced is the truth to our beginning.

Bottomline - is God Real or Not? Is God Dead, or is Evolution Dead? If you believe both options are true - that there is a God and *also* the hypothetical theory of evolution is true... this *cannot* be so as you will soon learn. Why? Well, for starters, they are diametrically *opposed* to each other in virtually every way possible.

"For the time will come when people will not put up with sound doctrine. Instead, to suit their *own* desires, they will gather around them a great number of teachers to say what their itching ears want to hear. They will turn their ears away from the truth and turn aside to myths." - *2 Timothy 4:3-4 (Bible)*

Who This Book is Written for

#1) THE CREATIONIST... This book is for those who already believe in God and want to share the *evidence for creation* with their family, friends, co-workers and even strangers. This book will teach you many great *evidences* for creationism that easily blow away evolution and that you can share personally with others. **Parents** *please* **read it and then make sure to share this with your kids and grandkids. I also hope you** *students* **from middle and high school through college share what you learn in this book with classmates, and even** *start creationism study groups* **to educate others and to fight back against evolution. The public school system obviously will** *not,* **so it's up to you and your friends.** (NOTE: One study shows that youth who were taught Biblical *creation* have a *much* greater chance of *not* falling into the evolution trap, especially when they leave home or go off to college).

CALL TO ACTION... I strongly suggest that you also give a copy of this book to everybody you want this information to reach. **Get a copy to** *influencers, too* (pastors, youth leaders, speakers, Christian bands, bloggers, social media gurus, etc.) so they can reach large groups of people fast. The *Eternal Time Clock* is ticking away. Tick-Tock.

#2) THE "NOT SURE" WHICH I AM, PERSON... This book is also for those of you who are just *not sure* what to believe, or haven't thought about it much, or may have come from a creationist background, but because of what you were taught in school may now be wondering if *"bacteria-to-people evolution"* could possibly be true, and you just want to know more about why you can believe *you were created for a reason,* and that you're *not* the great, great, great, great, etc., grandchild of a *sea slug* (I'm not talking about a lazy no good drunken sailor, either). **After reading this book, many of you will indeed lose any faith you may now have in the foolish** *theory of evolution.* Pretty obvious how much I detest the deceiving theories, hey? ;-) I'll try to tone it down a little more going forward. Maybe. OK, probably not.

#3) THE EVOLUTIONIST (ATHEIST/AGNOSTIC/SKEPTIC)... This book is also for the person who believes in some, or all of the theories of evolution (cosmic, chemical, stellar, organic, macro and micro). Please consider reading it with an OPEN MIND, *not* the typical, jaded evolutionary worldview of most evolutionists.

What This Book is About

This book is going to disturb some of you because of what you were taught in school about the farce of how every living thing evolved. It's time to stop the evolutionary propaganda, and for you to know the *other* side of the story, the amazing truth about *how* and *why* we were created by an all-intelligent God, and *not* by mere accident.

Keep in mind, I'm not twisting your arm, I'm simply presenting a strong case for *Intelligent Design.* **The decision is yours. Be open minded and you just may change your beliefs and potentially your eternity.**

BE AN AMBASSADOR... and order several copies of this book from CharlieWasWrong.com to share with other believers as well as unbelievers... such as family, friends, co-workers, pastors, youth pastors, students, and don't forget science teachers. Please share our website and FB posts often. People need to know this stuff.

In this book you'll learn a *lot* of proven and indisputable *evidence* why the kooky theories of evolution are actually just *unsubstantiated guesses.* Debunking the illogic theories is actually very *simple!!* And, it can even be a lot of fun in the process!!

***Cosmic Evolution* claims that the Universe started from Non-Intelligent, Random Chance... NOTHING.** Do you really think *nothing* created *everything?* Really? Come on now, that's nonsense... *think* about it.

The charade of ***Organic Evolution*** also claims that one single-celled microbe evolved from *nothing,* and then over eons of time, evolved into every plant, every animal and eventually every living human. It claims that **we are *all* related to each other.** (*Seriously, they say you're related somehow to a tomato. A fruit fly. A squirrel.* And, *every* other living thing). **They also *never explain how that FIRST living cell came to be.*** Trust me, they've tried really, really hard for a very long time. They can't explain it, and can't even reproduce a simple cell in their million dollar laboratories with thousands of scientists over the past 160 years or so, either. Guess why. It's *impossible,* that's why. Doesn't this alone bring a lot of questions to mind on the validity of *molecules-to-man* evolution? If not, it sure should.

In reality, ***Evolution is just a Fairytale for All Ages*** that has fooled much of the world since **the idea actually *originated* with the Greeks more than 2,000 years before Charlie Darwin was born.**

Darwin is *not* the *father of evolution* as many assume and preach in the *flimsy and unpredictable theory of evolution religion.* **He only *popularized*** it. In fact, there is much *proof* that **much of the content in his books was *plagiarized*** from others, including his own *grandfather.* Sadly, he devoted his life to a *proven-less fantasy* that has sadly lead multiple-millions of unsuspecting followers down this deceptive and madly destructive path for over 160 years.

NOTE: This book will provide (and further lead to) hundreds upon hundreds of powerful *scientific facts* and *mathematical probabilities* to *prove* that virtually all of the evolutionary assumptions about creation are **unscientific and *completely impossible.*** We can't say it enough… **Life did *not* and *can never* come from non-life!** It never has. Never will. Period. End of story. Still, many people will continue to believe and spread the assumptive *theory* of evolution over *intelligent* creation no matter how much evidence they are given to support creationism.

"Fools have no interest in understanding; they only want to air their own opinions." - *Proverbs 18:2*

The sad thing is… with *no evidence whatsoever,* even fools think they have the right answers about creation *(that we evolved starting from nothing),* and that Biblical Creation believers are the fools.

Don't be a fool. Rather, be open to explore this great debate of *Natural Evolution vs Supernatural Creation.* Start with this book.

Together, we'll explore a lot of solid *evidence* (not only Bible verses, but basic **common sense** and **an overwhelming abundance of real** *scientific facts* **and** *mathematical probabilities***) that support the Bible and prove that the foolhardy theory of evolution is nothing more than an abundance of colossal** *lies. Unscientific.* **A** *Grandiose Fairytale.* **Science** *Fiction.* **And,** *Completely Impossible. Not a fool, this Physicist and Nobel Prize winner said it all…*

"The pathetic thing about it is that many scientists are *trying to prove* **the doctrine of evolution, which** *no* **science can do."**
- Dr. Robert A. Milikan (Physicist and Nobel Prize winner)

SOME GREAT RESOURCES... Even though the information in this book alone virtually destroys all of the theories of evolution, it only scratches the surface of the information that's available... MUCH more can be found on *Creationism, Evolutionism, God, Christianity, Encouragement, Truth, Faith, Joy, Hope, Love, Apologetics, Atheism and related topics* - from these excellent resources: *EvolutionFacts.com, Creation.com, rzim.org, ICR.org, LivingWaters.com, AnswersInGenesis.org, CRU.org, Josh.org, DrDino.com, CreationToday.org, CBN.com, BillyGraham.org, Wretched.org, FCA.org, LeeStrobel.com, CrossExamined.org, KLOVE.com, Michael Behe, JohnLennox.org, Discovery.org CreationMoments.com, MetaxasTalk.com, EvolutionNews.org, NWCreation.net, ColdCaseChristianity.com, JmTour.com, ApologiaStudios.com, StephenCMeyer.org, BiblicalScienceInstitute.com, Alpha.org, HalLindsey.com, CharlieWasWrong.com,* **and the greatest resource of all - God's Word and Love Letter to us... the** *BIBLE (there's also a phone app called the YouVersion Bible, with audio and Bible studies, too).*

A Quick Overview of Evolution

In 1832 Scottish geologist *Charles Lyell* referred to gradual change over long periods of time. Charles Darwin only used the word in print once, in the closing paragraph of The Origin of Species (1859), and rather favored the phrases *"transmutation by means of natural selection"* and *"descent with modification."*

In the subsequent modern synthesis of evolution, Julian Huxley and others adopted the term, which thereby became the accepted technical term used by scientists. **Although in contemporary usage the term "evolution" most commonly refers to** *biological* **evolution, usage has evolved. The word also refers more generally to the "accumulation of change," which includes many disciplines besides biology.**

The Evolution of Evolution... The idea that life has evolved over time is *not* a recent one, and **Charles Darwin did** *not,* **in fact, come up with the idea of evolution** in general. For example, even **ancient Greek philosophers, like Aristotle, had ideas about biological development**. So, the theory itself evolved over short periods of time.

Origins of the Theory... The first significant step in the theory of evolution was made by *Carl Linnaeus.* His leading contribution to science was his creation of the binomial system of nomenclature - in lay terms, the two-part name given to species, such as *Homo sapiens* for humans. He, like other biologists of his time, believed in the fixity of the species, and in the *scala naturae*, or the *scale of life.*

Erasmus Darwin, **the grandfather of Charles Darwin,** was the first scientist to whom credit can be given for something starting to approach modern concepts of evolution, as noted in his contributions to botany and zoology. His writings contained many comments (mostly in footnotes and side writings) that suggested his beliefs in *common descent.* He concluded that vestigial organs (such as the appendix in humans) are leftovers from previous generations. The elder Darwin, however, offered *no* mechanism by which he believed evolution could occur. - *RationalWiki.org/wiki/Evolution*

Three Most Popular Definitions of So-Called Evolution

Some say the term *"evolution,"* **itself was actually invented…** by *Herbert Spencer.* Mr. Spencer introduced *sociology* into Europe, clothing it in evolutionary terms. He urged that the *unfit be eliminated,* so society could properly *evolve.* Others claim it goes back to the second century Greek writer Aelian (Aelianus Tacticus). Needless to say, it certainly was *not* Charles Darwin.

The Top Three Kinds of Evolution

Cosmic Evolution… The *BIG BANG.* Most evolutionists claim that the *universe* originated and evolved from absolutely *nothing,* by *random chance.* Meaning lots and *lots* of things just happened to fall in place all at the exact right time and place, even though the mathematical odds of creating space, matter and time from *nothing* are *zero,* as in the *Impossible Dream.*

Organic Evolution… Evolutionists claim that All *Living* Things originated and evolved from <u>*Non*</u>*-Living* Things.

Macro Evolution… Evolutionists claim that *animals* **and** *plants* **evolved from** *one species* **into** *others* (sea slugs became reptiles, reptiles became birds, apes became humans, etc.). All by *accidental chance evolutionary processes* over time, *every* time. Yeah, and monkeys fly (OK, maybe in the *Wizard of Oz,* but not normal everyday monkeys). Even **Mr. Darwin understood that species did** *not* **evolve, due to** *lack* **of any evidence.** However, because of some similarities, *he ignorantly concluded they must have all evolved from each other or a common ancestor.* Not science at all, only a faulty *theory.* Pseudo science. A bad guess. A fairytale. **To be considered actual science, it would need to have *empirical (observed) evidence* to support it. It does *not* have any such evidence.** Only made-up stories. And, of course key evolutionists will deny they are made-up stories, yet provide *no evidence* whatsoever.

Three Other Kinds of Evolution

Chemical Evolution... Says that *all* elements evolved from *hydrogen.* The first step of chemical reactions in the oceans in the development of life on this planet.

Stellar Evolution... Is the process by which a star changes (evolves) over the course of time. Some even claim that we came from stardust. Another wild *guess.*

Micro Evolution... Simply a *made-up name*, it's *not* evolution. There's *no* such thing as *micro-evolution.* **Changes within a true species are *not* evolution.** Micro-Evolution states that *variations* form within the *same* kind. (Dogs evolve into other kinds of *dogs*, birds staying *birds*, fish staying *fish,* etc). **This really is *not* Darwinian evolution** (i.e. *how evolutionists claim that one species randomly evolves into **another different kind** - such as cats becoming dogs*), but rather **it's known as *"adaptation,"*** which *all creationists agree on.*

Definition of Science

A Google search defines *science* as the intellectual and practical activity encompassing the systematic study of the structure and behavior of the physical and natural world through two key words... *observation* and *experiment.*

Two Kinds of Science

Observational Science... (Also known as *Empirical* Science) involves **things that can be** *observed* **and** *repeated* **through** *experimentation.* (Something evolution has *never* been able to accomplish, and we're *not* talking about *adaptation* within the *same* species).

Historical Science... alleged events that happened in the past and *aren't susceptible to scientific analysis* **because they** *can't* **be reproduced or repeated.** (ONLY the *fossil record* could prove evolution is true, and there are *no necessary transitional fossils* to be found anywhere in the world, as we have discussed).

Evolution is NOT Science
(Rather, it's a Religion)

"Evolution is a *Religion."* - *Michael Ruse (evolutionist and past professor of philosophy and zoology at the University of Guelph, Canada)*

The birdbrained *Theory of Evolution* **is actually** *not* science (i.e. must be *observed* in nature or the laboratory) as many people believe and are taught. **Rather it is simply the** *worldview* **of evolutionists.**

"My attempts to demonstrate evolution by an experiment carried on for more than 40 years have *completely failed.* **It is** *not even possible to make a 'caricature' of an evolution out of paleobiological* **facts... The idea of an evolution rests on** *pure belief."* - *Dr. Nils Heribert-Nilsson (Esteemed Swedish botanist and geneticist, of Lund University)*

Actually**, the fatuous theory of evolution is a *Religion*.** The *religion of naturalism* (atheism).

A dictionary **DEFINITION of RELIGION** is: *a pursuit or interest to which someone ascribes supreme importance.* Many will passionately *claim* that evolution is true science and vehemently argue that it's not a religion, but they are *incorrect.* Rather, **it *is* the definition of faith, of religion, and *not* observable and repeatable science.**

NOTE: **Evolution may be loosely referred to as *science* in parts of this book, but keep in mind, it is *not* science.**

Definition of Creationism

Creationism (Intelligent Design) - the belief that the universe and living organisms originate from specific acts of *divine* creation, as in the biblical account, rather than by natural *unguided, non-intelligent processes* such as evolution. **Prior to the mid-1800s, most scientists were *creationists*.** Most scientists are *evolutionists* nowadays, but really have no clue as to why they are, and certainly can't provide evidence for their beliefs. ***"Because I said so,"* is not a valid answer.**

Maybe your math and science teachers don't (didn't) believe that God created the universe, but it appears that Albert Einstein *did* indeed believe in *intelligent* design, and not *random chance* design.

"The mathematical precision of the universe reveals the mathematical mind of God." - *Albert Einstein*

Universal Common Descent

The imbecilic *Theory of Universal Common Descent* (all living organisms are *related* by descent from a common ancestor), is simply a pseudo-clever term for some evolutionists to try to unsuccessfully explain *molecules-to-man* evolution in a slightly different way than, *"we're just humans who evolved from monkeys."*

Basically, it says that *every one of us is related to... green alligators, long-neck geese, humpty-back camels and chimpanzees, cats and rats and elephants as sure as you're born... and don't you forget the unicorn.* Everybody sing...

They even go so far to say that **WE ARE ALL *RELATED* (as in distant *cousins*) to Apples, Oranges, Bananas and EVERY *PLANT* on Earth.** Come on, really guys? Fortunately, God gave us plants to enjoy and to help keep us alive, but we certainly *aren't* related to them. The stuff only silly fairytales are made of.

Evolutionary scientists can *say* whatever they want (and they do) and expect us to believe it, but there's a *huge insurmountable problem...*

There's absolutely NO *fossil record* or *evidence* to prove their wobbly *speculation* (Ah, I mean *theory)*. NONE. Just think about the big *multi-species reunion* someday in *imaginary evolution heaven.* The sad thing is, it won't be in the real heaven.

The Two Primary Worldviews

Theism: God is Personal, All Powerful, All Knowing, All Loving.
Atheism (Naturalism)**:** The rejection of God's existence.

Only These "Two" Options Explain Our Existence

RANDOM CHANCE EVOLUTION (Option #1)

A) **The Universe** (Cosmic Evolution) - Evolutionists believe (without any scientific evidence) that *nothing* somehow condensed by gravity *(wherever gravity came from)* into a single, tiny spot, and then decided to *explode into hydrogen and helium* (Sure it did. Fooled me. Or, maybe I should ask what *evidence* they have of this happening 13.8 *billion* years ago). This is best known as *Cosmic Evolution,* or what I like to call, the ***Bogus Big Bang theory.***

"Evolution is a *fairy tale* for grown-ups. This theory has helped *nothing* in the progress of science. It is *useless."* - *Bounoure, Le Monde et la Vie (October 1983) (Director of Research at the National Center of Scientific Research in France)*

B) **Life** (*Organic* and *Macro* Evolution) - Other than **adaptation** among species (i.e. changes among *like* kinds such as dogs becoming different looking dogs, yet *always* remaining dogs), which is scientifically proven and creationists also believe, in it, there is *no* (as in zero) scientific *evidence* for *Darwinian Evolution.* **More recently, *some* evolutionists now even agree that all animals came from *"like kinds"*** (cats from cats, horses from horses, birds from birds, etc), but **they have *no* answer to where these kinds *started and evolved from,* other than *matter* or *stardust*** *(that they also say came from NOTHING)* **evolving over billions of years.** To think people actually believe this nonsense even though there's *never* been a shred of real evidence, is beyond the scope of reality.

"The probability of life originating by *accident* is comparable to the probability of the unabridged dictionary resulting from an explosion in a printing shop." - *Dr. Edwin Conklin (Professor of Biology, Princeton University)*

INTELLIGENT DESIGN (Option #2)

The *only* other option is *divine intelligent design* (creationism) from a *supernatural* being. Let's just call Him, GOD. There are *no* other plausible options, other than the raving lunatics who say that *aliens* seeded planet Earth with life. They call it *panspermia*. These are the same loons who *can't* believe God did it all, but aliens did. Go figure. This of course begs the question, *"Where did these purported aliens come from?"* Meet George Jetson. Jane, his wife. I digress. Of course they never have any actual *proof,* we're just supposed to blindly believe them. Why? **Isn't that the epitome of *blind faith?*** *"Yeah Nick, there's no God, but let me tell you about little green spacemen who actually started life on Planet Earth."*

*This Nobel Prize Winner, Harvard Professor **AND** Evolutionist says...* "Life Arose as a *Supernatural* Creative Act of God."

"When it comes to the origin of life, we have *only* two possibilities as to how life arose. One is *spontaneous generation* arising to evolution; the other is a *supernatural creative act of God.* There is *no* third possibility... Spontaneous Generation was scientifically *disproved* over one hundred years ago by Louis Pasteur, Spellanzani, Reddy and others. That leads us scientifically to *only* one possible conclusion – that *Life Arose as a Supernatural Creative Act of God* ... I will *not* accept that philosophically because I do *not* want to believe in God. Therefore, I choose to believe in that which I know is scientifically *impossible,* spontaneous generation arising to evolution." - *Dr. George Wald*

(Evolutionist and Professor Emeritus of Biology at the University at Harvard, Nobel Prize winner in Biology).

See, no matter how convincing the evidence is supporting creation, and against evolution, many people will still *not* choose the truth, and reject God in exchange for the lies of evolution.

OK, now let's get into this stuff a little deeper. I think you'll find it very interesting. Each topic is kept insightful, yet as brief as possible simply so you will read them. You can always dig much deeper starting at **CharlieWasWrong.com** and our *Facebook* page **Facebook.com/CharlieWasWrong** (If you're a creationist, please join us daily and share it).

Beware of Evolution's *Talking Heads*
And, the Theory of..."Monkey Science"

Modern day evolution *Talking Heads* (leaders of their fragile and witless theory) spew such silly stuff about creation, trying relentlessly to throw a monkey wrench into the mix. I like to call what they preach from their Darwinian religion, *"Monkey Science."* Because that's exactly what it is. Or, maybe we should call it *"Simple Science."* Yeah, I like that. How about you? Well, actually it's *not* science at all but please allow me to have some fun here.

Many speak with eloquence and confidence and may even sound intelligent about evolution at times, but what comes out of their mouths is usually *low-level, unproven* and *incorrect pseudo-science* that they want you to believe is superior to creationism. It's not.

They'll use these *tactics* to try to distract creationists from the important topics that they themselves *can't* answer. Don't allow

them to do this. They'll bash creation science all day long as if evolution is true and intelligent design is false. Wrong again.

They'll talk about the first cell forming like they were there and that it's actually, mathematically possible, which it's *not.* They'll respond something like, *"How did the first cell form? Huh. Why do you gotta ask me that? Who cares, it just formed. You know, in a slimy mud hole or something like that. Nobody really knows exactly how and nobody can prove it, but who really cares anyway? Trust me, it just formed somehow, everybody knows that."* (Oh gee, that explains it then. I guess writing this book is a big waste of my random time. Maybe just for kicks we'll talk about the real science behind the *scientific impossibilities* and *zero mathematical probabilities* of all this having any remote chance of ever happening. Then, let's compare notes again at the *end* of this book, why don't we).

And, they'll talk on and on about all kinds of thoughtless theory of evolution stuff with *Monkey Science* explanations. *"Yep, we all descended first from a single cell bacteria climbing out of a mud pit, which went on to find the opposite sex bacteria that just so happened to come out of the same pit at the same time to create something beautiful together, like another ugly bacteria… and, that eventually even became plants and apes and such (or the same common ancestors), so yeah, we are all actually related. You, me, my dog Louie, your goldfish, and yes even the juicy grass fed steak I ate last night. Heck, I guess we're all even related to your book Nick since it did start out as a tree, right?"* (That's special, let's all sing Kumbaya. That's how crazy and weird all this evolutionism stuff gets).

This book will go into over 150 different topics. It will challenge you to *think*. To *dig deeper*. And, *challenge evolution's baseless claims and unprovable stories.* Is that fair game? Of course it is.

If you listen *closely* and *compare* what they're saying and talking about to the much more detailed and true, creation science, you'll see that **evolution is *extremely simple, surface and silly science.* OK, like I said earlier, it's *not* really science. It's actually a *faith,* a *religion, pseudo science,* and an *unproven hypothesis.* It's a *Guessing Game* where they *change the answers all the time to try to fit their worldview.*** Evolutionists hate when you call evolution a faith or religion but that's *exactly* what it is. It's a far cry from actual science.

"The more one *studies* paleontology, the more certain one becomes that evolution is based on *faith* alone... exactly the same sort of faith which it is necessary to have when one encounters the great mysteries of *religion.*" *- Louis Trenchark More (Evolutionist. Quoted in Science and the Two-tailed Dinosaur, p. 33)*

Keep in mind creationist friends... many (certainly not all) atheists think they have the answers to our creation when it comes to science, and that *you* are pretty much an idiot when it comes to this topic. They may not tell you that to your face, but many are thinking it. My experience is that most of them will *not* really listen to you very close, and think whatever you say is "wrong," or "stupid."

Before they try to push their **Monkey Science** agenda down your throat any longer *(many times by asking you ridiculous questions)...* **beat them at their own game by telling them that you will *not* answer their questions until they *first* answer the more important questions** such as, *"How did nothing become something"* (actually every living thing in the universe), and *"Where is the fossil record (evidence) for macro-evolution?"* (Many other questions are also listed at the end of this book. They *won't* be able to *correctly* answer these so don't worry).

Did the Universe *Explode* into Existence? (by a *Random Accident from Absolutely NOTHING... OR was it SUPERNATURALLY CREATED?)*

The Really BIG Bang Theory
(A Universe from NOTHING?)

The *Big Bang Theory* is almost as crazy to believe in as the *Flying Spaghetti Monster*. Both are elaborate fairytales and both are *scientifically and mathematically impossible.*

Come on now, admit it… deep down, do you really believe what the absurd evolutionary view of *cosmic evolution* teaches - which is that *"NOTHING"* created the entire universe, and continued on to create every living thing we see, hear, touch and smell around us every day? That somehow this all *magically evolved* out of *non-intelligent* thin air… all by complete *random, purposeless chance,* nevertheless. REALLY?

And, it *had* to all start from *nothing,* otherwise if it did start from *something (a tiny speck of whatever),* **we need to ask… "Where did that *tiny something* come from?"** It certainly *wasn't* eternal. *Scientists agree that the world could not possibly be "eternal" or it would have literally burned-up by now,* due to the fact of the immutable *Second Law of Thermodynamics,* where energy is winding down and matter left alone tends toward chaos and randomness, *not* greater organization and complexity as would be *required* for evolution to be true.

This is really crazy stuff friends. **THINK for yourself. Don't fall for the** *mindless evolutionary stories* **you've been told and sold.** Intentional or not, they are *half lies, total lies,* and *total impossibilities.*

Use your God given *Common Sense...* **No matter how hard they sell it and how believable it may sound, the universe (including the Earth, Sun and Moon) and every living thing (plants, animals and humans) did** *not* **and could** *not* **evolve from** *NOTHING.*

THE ODDS of the BIG BANG Actually Happening...

"It is *no* **more likely that our world has evolved out of chaos than that a hurricane, blowing through a junkyard, should create a Boeing 747."** - *Sir Fred Hoyle (Non-Creationist and famous astronomer and mathematician who formulated the theory of stellar nucleosynthesis)*

The universe being made by a *supernatural God* is more logical than the Big Bang Theory and deep down you know it. *It takes much more faith to believe in evolution, than it does to believe in God.*

The *Big Bang* **"fairytale" theory was** *invented* **in 1927** and basically says that the universe started from *nothing* somewhere around 13.8 billion years ago. How are they so sure it wasn't 12.4 billion, or maybe 14.2 billion years, 3 weeks, 2 days, 27 minutes and 7 seconds ago? Lol. **Evolution believing scientists act like they have all the answers about creation, but in reality they really** *don't* **have a clue.**

The rickety evolution assumption then goes on to say that this *nothingness* **somehow condensed by gravity into a small spot, and decided to somehow explode.** This turned into gas clouds that then somehow mysteriously became stars over time. Scientists make some seemingly bold and convincing claims, but of course **the inventors of this wild theory have** *nothing* **to actually** *prove* **their big bang**

hypotheses. They are made up stories! That's all. Apparently, we're all supposed to simply go along with their erroneous theories. Not me.

PICK <u>ONE</u>… **Either** *God is Eternal* **OR** *Mass Energy is Eternal.* It *can't* be both. (The *immutable Second Law of Thermodynamics* proves that *Mass Energy* "**cannot**" **be eternal** or none of us would be here). There is only *one* choice. Choose wisely. **The challenge to you is to** *challenge* **this shaky house-of-cards theory, and not just gullibly believe and accept it.**

"In the beginning, GOD created the Heavens and the Earth."
- *Genesis 1:1*

Several Reasons That Completely *Demolish* the Big Bang Theory

If just ONE of these reasons below is true, and they are *all* true by-the-way, then *the infamous Big Bang never happened.*

Simply stated… **The Impossible Big Bang theory stands in clear** *violation of physical laws, celestial mechanics, and common sense.*

Following are a number of scientific reasons why the erratic Big Bang theory is *unworkable* and *fallacious.* Big Bang go BOOM!!

1 - **The evolutionary Big Bang theory is based on** *theoretical extremes.* It may look good in math calculations, but it *couldn't* actually happen. A tiny bit of *nothing* packed so tightly together that it blew up and produced all the matter in the universe. Seriously now, this is nothing more than a *grand scale fairytale.* It's a bunch of armchair calculations, and *nothing* more. *It's easy to theorize that something is true* when it has *never* been seen and there is *no definitive*

evidence that it exists or ever happened. **Let's not mistake** *Alice in Wonderland theories* **for real science.** BOOM!! (or should I say BANG?)

2 - *Nothingness* **CAN'T pack together.** It would have *no* way to push itself into a pile. BOOM!!

3 - **A vacuum has** *zero* **density.** It is said by cosmic evolutionary scientists that the *nothingness* got very dense, and that's why it exploded. However, **a total vacuum is the** *opposite* **of total density.** (How *nothingness* can do anything is beyond the scope of reality, but *apparently not science fiction).* BOOM! !

4 - **There would be** *no ignition* **to explode nothingness.** No fire and no match. It could *not* be a chemical explosion, for *no* chemicals existed. It could *not* be a nuclear explosion, for there were *no* atoms! BOOM! The Big Bang is *not looking so good* so far, but there's more.

5 - **There is** *no* **way to** *expand* **it.** How can you expand what *isn't* there? Even if that magical vacuum could somehow be pulled together by *gravity* (which apparently also magically appeared on the scene by *random chance processes,* just in time nevertheless), what would then cause the pile of *emptiness* to push outward? **The** *gravity* **which brought it together would keep it from expanding.** BOOM!!

6 - **Nothingness can<u>not</u> produce heat.** The intense heat caused by the *exploding nothingness* is said to have changed the nothingness into protons, neutrons, and electrons. *First,* an empty vacuum in the extreme cold of outer space *cannot* get hot by itself. *Second,* an empty void *cannot* magically change itself into matter. *Third,* there can be *no heat* without an energy source. *Forth...* BOOM!! ;-)

7 – The calculations are *too* **exacting.** Too *perfect* an explosion would be *required*. On many points, the theoretical mathematical calculations needed to turn a Big Bang into stars and our planet *cannot* be worked out. In other words, they are too *exacting*. Knowledgeable scientists call them *"too perfect."* Mathematical limitations would have to be met which would be *impossible* to achieve. The limits for success are simply far too narrow. BOOM!!

8 - Such an equation would have produced NOT a universe, but a HOLE. Even *evolutionist* Roger L. St. Peter in 1974 developed a complicated mathematical equation that showed that **the theorized Big Bang could** *not* **have exploded outward into hydrogen and helium.** In reality, St. Peter says the theoretical explosion (*if* one could possibly take place) would fall back on itself and make a theoretical black hole! *This means that one imaginary object would swallow another one!* BOOM!!

9 - There is *not* **enough** *antimatter* **in the universe.** This is a *big* problem for the theorists. The original Big Bang would have produced equal amounts of *positive matter* (matter) and *negative matter* (anti-matter). *Only small amounts of antimatter exist.* There should be as much antimatter as matter - if the Big Bang was true. BOOM!!

10 - The antimatter from the Big Bang would have *destroyed* **all the regular matter.** This fact is *well-known* to physicists. As soon as the two are produced in the laboratory, they instantly come together and *annihilate* one another. BOOM, BOOM!! - *EvolutionFacts.com*

We could go into *many* more scientific facts about this topic but the above should be enough information to give you a good idea that **our universe did** *not* **and** *could never* **possibly happen by** *pure luck* **or** *helter-skelter evolutionary chance.* Stop believing this harebrained, disproven myth of the big bang. It's only another fairytale.

The universe; including the Sun, Moon, planets, stars, Earth and everything on and in it could *only* have been created by a *supernatural* being. He is very real and His name is GOD and if you don't know him yet, you would do well to get to know Him. Maybe today is that day. Seriously, if you're a non-believer… stop the anti-God madness. He has sooooo much more to offer you than the un-credible theory of evolution. Do yourself a big favor and learn all you can about Him.

Again, I hope you're reading this with an open mind. It's OK to question these topics, you can always *explore* what you read here in greater detail in many places online or off. In fact, you should. **When they are among the most important questions in life** *(where you came from and where you'll spend eternity)*… **I hope you believe that seeking the *truth* is of ultimate importance.**

"'Scientists who go about teaching that evolution is a fact of life are great *con-men*, and the story they are telling may be the *greatest hoax ever*. In explaining evolution - we do *not* have one iota of *fact.*' [Tahmisian called it] a tangled mishmash of guessing games and figure juggling." - *Fresno Bee (August 20, 1959, p. 1-B, quoting evolutionist T.N. Tahmisian, with the Atomic Energy Comm)*

BOTTOMLINE… **There was NO Big Bang!!** *Something* can **NEVER** come from *Nothing. Something* (i.e. *the Universe and all that is in it*) came from *Something,* and that *Something* was Almighty GOD. Put your trust in *Him,* not in *chance.*

Our Expanding Universe

In 1929, astronomer Edwin Hubble (Hubble telescope inventor) found that **galaxies are moving *away* from each other. Meaning that *the universe is expanding.***

Among modern day skeptics and new atheists ("New" Atheists is a term coined in 2006 for modern day atheists), one of their well known so-called gurus, Lawrence Krauss, a physicist/cosmologist, has said that because of Hubble's discovery, science can explain why the universe is expanding, without needing the Bible. What he fails to recognize is the fact that **the Bible mentions several times that the universe is *expanding,* 3,000 years *before* the Hubble telescope was invented.**

This, often disrespectful atheist, goes on to say, *"We all came from stars that exploded billions of light years away. So forget Jesus, the 'stars' died so that you could be here today." - Creation.com*

Yeah, and a genie in a bottle just gave me 3 wishes. *There, I said it - so you should believe it.* **So, this dude is a leading evolutionist and this is the best he's got? Seriously?? Come on now, that's the best of his best stuff about evolution?** If you can believe his incredibly weak blather over the tons of strong evidence *against* evolution in this book and many other resources available to you, then **take a look at your *heart,* because it's apparently taken control over your *mind.***

As if he, or anybody for that matter, has any kind of clue about this whatsoever. **It's all *100% speculation,* nothing more. Self proclaimed experts like this often times say whatever they want and millions of people blindly believe them. Don't make that grave mistake. Think for yourself. Do your research if you have doubts about *stupid claims* made by people like this.** They are good at marketing and branding *themselves,* I'll give them that. It may stroke their egos and fill their bank accounts from book sales and speaking engagements, but **just making a statement about something *doesn't* make it true.** Always keep that in mind.

As I said before and will probably say again, in this book **we'll debunk utter nonsense like this - with** *real scientific facts,* *mathematical statistics,* **and a** *thinking* **person's** *common sense.*

"The universe and the Laws of Physics seem to have been specifically *designed* **for us. If any** *one* **of about 40 physical qualities had more than slightly different values, life as we know it could** *not* **exist: Either atoms would not be stable, or they wouldn't combine into molecules, or the stars wouldn't form heavier elements, or the universe would collapse before life could develop, and so on..."** - *Stephen Hawking (Atheist and best known scientist since Albert Einstein)*

Solar Collapse

Research indicates that our *sun* is gradually shrinking at a steady rate of seconds of arc per century. **At its rate of shrinkage, as little as 50,000 years ago the sun would have been so large that our oceans would boil. But in** *far less time* **than 50,000 years,** *life here would have ceased to exist.* Recent studies have disclosed that neither the size of the sun, nor our distance from it, could be much greater or smaller - in order for life to be sustained on our planet.

Extrapolating back 100,000 years ago, the sun would have been about twice its present size, making life untenable. - *EvolutionFacts.com*

In other words, since evolutionists claim the sun has been around for billions of years... their theory (assumption) is just a fairytale, and nothing more.

Why haven't aliens come to our solar system?
They checked the reviews. One star.   :-)

The Immutable "First" Law of Thermodynamics

LORD KELVIN *(Thermodynamics and Energetics. Physics)...* **"Do not be afraid to be free thinkers. If you think strongly enough, you will be forced by science to the belief in God."** - *"The atheistic idea is so nonsensical that I cannot put it into words."* (Lord Kelvin, Vict. Inst., 124, p. 267, as cited in Bowden 1982, 218)

EVOLUTION TEACHES... that *matter* is *not* conservative but rather its *self-originating;* it can arise from nothing and increase. *The First Law of Thermodynamics* teaches the *OPPOSITE* and thus *annihilates* this huge error.

The First Law of Thermodynamics is immutable *(unchangeable)* and says that... **Energy cannot by itself be created nor destroyed. Energy may be changed from one form into another, but the total amount remains *unchanged.***

Einstein showed that **matter is simply another form of energy,** as expressed in the equation: $E = MC_2$.

According to the First Law, the sum total of energy (or its twin sister, *matter*) will always remain the same. None of it will disappear by itself. (The corollary is that *no* new matter or energy will make itself. **Thus, the universe could *not* be created from *nothing*** as so-called *Cosmic Evolutionists* try to sell us) - *EvolutionFacts.com*

"Thus the Heavens and the Earth were finished, and all the host of them." - *Genesis 2:1*

The Immutable "Second" Law of Thermodynamics - *(the Exact "Opposite" of Evolution)*

EVOLUTION TEACHES... that matter and living things keep becoming *more* complex and continually evolve toward greater *perfection.* Just as inorganic matter becomes successively more ordered and perfect *(via the Big Bang and Stellar Evolution),* so living creatures are always evolving into higher planes of existence *(via Species Evolution).*

The Second Law of Thermodynamics devastates the sleepy, worn out theories of evolution.

A law is a principle that is *never,* ever violated. A huge problem, and one that makes evolution *impossible,* is the *Second Law of Thermodynamics,* also called the *Law of Increasing Entropy* (increasing *disorder*).

This *unchangeable law*... states that energy is winding down and matter left alone tends toward *chaos* and *randomness, NOT* greater organization and complexity.

On the other hand...

Evolution requires the exact *opposite* process, which is *not* and has *never* been observed *anywhere* in nature, ever. No, no, never, never, uh uh uh. Thus, making the Big Bang... a Big Blunder, a Big Blooper, a Big Bad Bomb, a Big Botched Boo Boo, etc.

The point is, the Big Bang could *not* have possibly happened. *So if you subscribe to that silliness, you can stop right now, this minute.* ;-)

Don't try telling me that a *tornado* debunks this law. Nobody is buying that line because it's nonsensical rubbish. Some evolutionists actually believe it and use that line to support the law of *Increasing Entropy.*

This unchangeable law *proves* that *Cosmic Evolution* is false (it did *not* and it *cannot* happen). **If the Earth is billions of years old, it would have *burned up* and *ceased to exist* long ago.**
- EvolutionFacts.com

It's funny how **the only science that *disagrees* with the Second Law of Thermodynamics is *Evolutionary Biology.*** It destroys their worldview and all they stand for about creation. Evolutionists can believe and say whatever they want, however, this LAW of Physics is *Immutable* (unchangeable). The *unstable, counterfeit theory of "evilution"* busted in the chops once again, goes down for the count.

SCORE: Intelligent Design = 102. Evolution = - 2.

"Atheism is so Senseless. When I look at the solar system, I see the Earth at the right distance from the sun to receive the proper amounts of heat and light. This did NOT happen by *chance*. The true God is a living, intelligent, and powerful being." - *Sir Isaac Newton (Father of Modern Science)*

The Law of Manufacture

"The *evolution theory* is purely the product of the *imagination.*"
- Sir Dr. Ambrose Flemming (Father of modern electronics)

We could word the *Law of Manufacture* something like this: *"The maker of a product has to be more complicated than the product."* The equipment needed to make a bolt and nut has to be far more complex than the bolt and nut. This is the *First Law of Products.*

Here is another *law* to consider. We will call this one the **Law of Originator,** and describe it in this way: ***"The designer of a product has to be more intelligent than the product."*** I hope you would agree. There are *no* exceptions. See where we're going with this?

Let's return to the bolt and nut for our example of what we will call our *Second Law of Products.* Neither the bolt nor the nut made themselves. In Addition: **The person who made this bolt and nut had to be *far more intelligent* than the bolt and nut, and *far more intelligent* than the production methods used to make them.**

Do you still think the universe created itself? That all living things created themselves without a Master Designer? These are *much* more complex and dependent on each other for survival than nuts and bolts. - *EvolutionFacts.com*

Think hard about the Bible verse below. Read it a few times...

"The god of this world (satan) has *blinded the minds* of the unbelieving so that they might *not* see the light of the gospel of the glory of Christ, who is the image of God." - *2 Corin. 4:4*

An Eternal Universe?

To try to get around insurmountable difficulties such as *natural selection* and *mutations,* some evolutionists are now postulating that the universe is *eternal,* because if time is eternal, they argue, then theoretically, given enough time any event can happen. Nonsense. Sounds like a fairytale because it's absolute hogwash. (**Evolutionists love to use the phrase, *"given enough time, anything can happen."***) Again, **they still need to provide *evidence*.** Words alone in this case are meaningless and without merit.

The idea of an eternal universe *cannot* be substantiated, however, because the universe is slowly approaching *"heat death"* in accordance with the *Second Law of Thermodynamics*. Heat death will occur when all the energy of the cosmos has been degraded to random heat energy, with random motions of molecules and uniform low-level temperatures. If the universe is eternal, this state would have been reached *a long time ago*. **The fact that the universe is not dead is clear *evidence* that it is *not* infinitely old.**

What doesn't make sense is that many evolutionists can somehow believe that the universe is *eternal,* yet they refuse to believe there can be an *Eternal Creator,* an intelligent mind behind it all. On the surface, that makes *no* sense, but maybe for some, there is more to their blind faith than meets the eye. *- EvolutionFacts.com*

The Bible may make more sense of it for you...

"The wrath of God is being revealed from heaven against all the godlessness and wickedness of people, who suppress the truth by their wickedness, since what may be known about God is plain to them, because God has made it plain to them. For since the creation of the world God's invisible qualities - his eternal power and divine nature - have been clearly seen, being understood from what has been made, so that people are without excuse. For although they knew God, they neither glorified him as God nor gave thanks to him, but their thinking became futile and their foolish hearts were darkened. Although they claimed to be wise, they became fools." - *Romans 1:18-22 (The Bible)*

Strong Evidence for a Master Designer Comes From a... *Fine-Tuned Universe*

- The *electromagnetic coupling constant* binds electrons to protons in atoms. If it was smaller, fewer electrons could be held. If it was larger, electrons would be held too tightly to bond with other atoms.

- Ratio of electron to proton mass (1:1836). Again, if this was larger or smaller, molecules could not form.   A coincidence? No way.

- Carbon and oxygen nuclei have finely tuned energy levels.

- Electromagnetic and gravitational forces are finely tuned, so the right kind of star can be stable.

- Our sun is the right color. If it was redder or bluer, photosynthetic response would be weaker.

- Our sun is also the right mass. If it was larger, its brightness would change too quickly and there would be too much high energy radiation. If it was smaller, the range of planetary distances able to support life would be too narrow; the right distance would be so close to the star that tidal forces would disrupt the planet's rotational period. UV radiation would also be inadequate for photosynthesis.

- The Earth's distance from the Sun is crucial for a stable water cycle. Too far away, and most water would freeze; too close and most water would boil.

- The Earth's gravity, axial tilt, rotation period, magnetic field, crust thickness, oxygen/nitrogen ratio, carbon dioxide, water vapor and ozone levels are just right. - *AnswersinGenesis.org*

Former atheist *Sir Fred Hoyle* said, **"Commonsense interpretation of the facts is that a *super-intelligence* has monkeyed with physics, as well as chemistry and biology, and that there are *no* blind forces in nature."**

The Sun... *(Another Evolution Slayer)*

Just the SUN alone completely *CRUSHES* the *faulty theory* of random chance *Stellar Evolution*. Here is where not only *science* and *mathematical probabilities* come into play, but *common sense* as well.

Our *Sun* is *rotating* rather *slowly,* but the *planets are rotating far too fast* in comparison with the Sun. In addition, they are *orbiting* the Sun far faster than the Sun is itself turning. But if the planets did *not orbit* so fast, they would hurl into the Sun; and if the sun did *not rotate slowly,* it would fling its mass outward into space.

According to *David Layzer* of Harvard *(a non-creationist),* in order for the Sun to originally have been part of the same mass as the planets and moons, it would have to rotate *ten-million times faster.* Layzer adds, if the Sun lost so much of its momentum, why did the planets not lose theirs? Why indeed.

If all the planets originally spun off the sun as evolutionists say, why are Uranus and Venus rotating *backward,* compared to all the other planets? The other seven rotate forward, in relation to their orbit around the Sun. *- EvolutionFacts.com*

Did you ever think to yourself why the Sun is the *exact size* and *exact temperature* it is, and why it's the *exact distance* from Earth it needs to be, and **why it *continues* to *stay* virtually this exact same**

size, temperature, and distance (about 93 million miles) **from Earth... so all living things can sustain life? Not a random act.**

Also, why is the gravitational force *exactly right* so as not to suck the Earth up into it? And, how could *gravity* itself have just happened by *random chance,* like the millions of other things evolution teaches that *just happened?*

And, *how* **and** *why* **does the Sun even *exist?* And, *why* do humans need the Sun to exist in the first place.** Hey, it's not just for tanning.

If ANY one of these things were *slightly* different, planet Earth would cease to exist.

If you think this all started and then continues to happen by *random, unplanned chance (day after day)* and not by the *purposeful intelligence* of an *all powerful Creator,* **you are being suckered into believing some of the biggest science *fiction* fantasies of all time.**

Come on friends... Think hard about it - this stuff can NOT happen by *random chance!!* Period!! It's totally impossible!

Stop believing this utter nonsense (if you do), and start believing in the God who created you.

Bottomline... *How* did the Sun come to exist in the first place, and *how* did it know it was necessary for life? Evolution does *not* have any answers.

Find out *why* you're alive today and can actually understand what you're reading. Why you can love, laugh, cry, reason, and have emotions. It's *no accident* this all happens in all of us. Rather, it's *divine design.* Once you understand and get to know God, life becomes *meaningful* and *amazing!!* You have *HOPE.* You're *not* reading this book by some *random accident.* (More on this to come).

The Moon *(God's Little Light)*

Pretty much the same goes for the Moon as well as the Sun. The Moon does NOT exist by *random, designer-less chance,* friends. Never in a billion years. Or even 4.5 billion for that matter.

The Earth's gravity keeps the Moon in orbit. It's the *exact* distance from Earth that it needs to be. Think these things just happen by aimless *chance?* Think again. No way Jose. As the Moon orbits the spinning Earth, there is a cycle of two high tides and two low tides about every 25 hours. Tides are vital to life on Earth. Tides cleanse the ocean's shorelines, and help keep the ocean currents circulating, preventing the ocean from stagnating. - *Creation.com*

Every moon in the universe is located at the *precise distance* to keep it from flying into or away from its planet. How could all this originate from a single explosion or collision? Or many? Of course it *couldn't* have. *NONE* **of the natural grandiose theories fit into the laws of physics.** Stop being fooled by evolutionary theories.

I could go on and on with many of the topics in this book but I don't want it to be 10,000 pages and neither do you. You can find much more information on this and many other topics that pulverize evolution, at the other great references mentioned earlier in this book.

The Earth *(Where the Magic Happens)*

So, the Earth somehow magically formed several billion years after the so-called Big Bang, where supposedly *nothing* exploded by *accident* almost 14 billion years ago *they* say? **The Moon *just happens* to orbit the Earth** *(by random chance)* **and the Earth** *just happens* **to orbit the Sun** *(again, by yet another freak random chance)?* **THINK about this friends. There is absolutely NO way this happened/happens by *purposeless, unguided chance!!*** This *alone* should give you serious doubts about the impossibility of the Big Bang and all evolution theories.

All plants, animals and humans on Earth *just happened* to live on the *only planet in the galaxy to sustain human life*, with things evolution tells us again that *just happened* by *random chance;* like the Sun, the Moon, oxygen, carbon dioxide, water, food, gravity, coffee, bacon, etc? *(**If God didn't create all of this, can someone please explain coffee and bacon to me? There's no way these two evolved**).*

Man *and* plants would need to be *"fully*-formed" and living in the same geographical area to sustain life from the very beginning. You know, the human breathes out *carbon dioxide* for the plant to live, and the plant returns the favor with life-breathing *oxygen*. And, then there's the Sun, water, and food also needed to maintain life. Furthermore, all the parts within our bodies need to assimilate oxygen, water and food. Evolution is *not, Not, NOT* your answer. Be wise and stop believing in it, if you do. Go away evolution, you bother us.

Atheists want to deny there's a God, but need an *alternative* solution. Of course that's why, when examined closely there's *no* credible evidence. Even though they sell it hard, the dubious theory of evolution is *devolving* and *dissolving* more and more each year.

Oh, but how could we possibly believe in God? Maybe because **as we look closer, we discover that the** *unfounded evolutionary theory* **is an** *unusually demented and preposterous concept,* **and altogether** *far-fetched and archaic.* **If just** *one* **of these life essentials (Earth, Sun, Moon, oxygen, blood, water, food, etc.) was not in the mankind story exactly as it is now, we all would cease to exist. The mathematical chances of all (or any) of these** *evolving* **to sustain life is <u>ZERO</u>!!**

"Evolution is Religion."

Michael Ruse
*(Evolutionist and past professor of philosophy
and zoology at Guelph University)*

From NOTHING... to Rocks and Slime... to a Living Cell... to Fish... Reptiles... Birds, Monkeys... and finally to Humans. Oh REALLY?

Let's examine the *so-called evidence*
for evolution, or complete lack thereof

CREATION speaks loudly to those who are actually *listening*. Are *you* listening? If not, maybe it's time you do.

Check this out... Charles Darwin did *not* know of even *one* instance in which *any* species changed into *another* species. The same holds true to this day.

"Not *one* change of species into another is on record... we *cannot* prove that a single species has been changed." - *Charles Darwin (My Life and Letters)*

If that quote from *Mr Evolution* himself didn't sink in, *read it again.* **I'll bet most of you were *not* aware of this quote he made because that has been *hidden from you* either because it was *not* known by the presenter, or *intentionally hidden* by those who choose *not* to share it** - such as; the high school and college textbooks, pseudo science magazines, TV documentaries, your science teachers, fake news, etc. Keep in mind that **there is *a whole lot more* than this that they *also fail to mention,* in order to continue to propagate this**

deceptive worldview. Even if you're a non-creationist, doesn't this bother you somewhat? If you're seeking *truth*, it should.

And, **absolutely** *NO* **evolution from one species to another has** *ever* **occurred even** *once* in over 160 years since then!! If it had, there would be more than plenty of *evidence* to substantiate the bogus evolutionary claims and stories. News would be everywhere, and I would *not* have written this book. This is a monster clue in itself.

Spontaneous Generation *(Life From Non-Life, Evolution. AKA: Abiogenesis or Biopoiesis -* is Absolutely 100% Completely... <u>IMPOSSIBLE</u>!!

"Life could *not* **arise spontaneously in a primeval soup of any kind.... Furthermore,** *no geological evidence* **indicates an organic soup** *ever* **existed on this planet. We may therefore with fairness call this scenario the** *'myth of the prebiotic soup.'"* - *Sir Fred Hoyle (Famous Astronomer who formulated the theory of stellar nucleosynthesis. Also, an ex-atheist)*

SPONTANEOUS GENERATION (Definition)... **the** *"hypothetical"* **process by which living organisms develop from nonliving matter; also, the** *archaic* **theory that utilized this process to explain the origin of** *life.* According to that theory, pieces of *cheese* and *bread* wrapped in rags and left in a dark corner, for example, were thus thought to produce *mice,* because after several weeks there were mice in the rags. Many believed in spontaneous generation because it explained such occurrences as the appearance of maggots on decaying meat. **-** ***Encyclopedia Britannica*** *(Note: They actually said "hypothetical" and "archaic" in the description)*

It doesn't matter that some scientists *make up big words* like *Spontaneous Generation* and *claim* that is how we all came to exist… Many times the *words* people say are *meaningless,* especially when it comes to the *existence of mankind.* They say just about *anything* they want and get away with it. Some wildly creative stuff at that.

It's important to always keep in mind that **ONLY the *EVIDENCE*** (or *lack* thereof) **really matters and can tell us the *real truth.*** It can't be said enough… *the delusive theory of evolution has no evidence and is a flat-out sham, one of the biggest frauds ever orchestrated against mankind. It must be stopped as soon as possible, and you can help.*

"It is no more likely that our world has evolved out of chaos *(random chance evolution)* **than that a hurricane blowing through a junkyard should create a Boeing 747."** - *Sir Fred Hoyle (Famous Astronomer and former atheist)*

Louis Pasteur *Proved that* Life <u>ALWAYS</u> Comes From LIFE. One of the greatest scientists ever, *Louis Pasteur,* made a major breakthrough discovery in the 1860s that **confirmed *Biogenesis,* meaning…** *life <u>always</u> comes from life.* Thus, Pasteur *proved* without a shadow of a doubt, that *Spontaneous Generation* (life supposedly arising from *non-*life) is *IMPOSSIBLE.*

It's an immutable (unchangeable) law… It has never, and will never be proven different. Pasteur concluded from his experiment that *only* God could create living creatures. But modern evolutionary theory continues to be based on this *outdated assumption* **disproved by Pasteur over 150 years ago.** They act like it was just last year.

INSTANT SUCCESS NECESSARY… In order for life to arise from non-life, there would have to be *INSTANT SUCCESS.* **ALL of the parts** (i.e. **heart, lungs, blood, central nervous system, brain,**

skin, etc) would suddenly *have* to be there... and *all* would have to *immediately* function with essential *perfection,* or there would be <u>*NO LIFE*</u>.

FOR LIFE TO OCCUR... *DNA, protein, amino acids, enzymes, and all the rest* required to form the *first living creature* (*every* living creature for that matter) - would *have* to link up with ease into long, extremely complicated coded strings. **In addition,** *thousands* **of other complicated chemical combinations would** *have* **to be accomplished all within a** *very short time of life forming.* How long could you live without a beating heart? How long without blood, veins and arteries? Without lungs? Without oxygen? The list goes on and on. **They must all be there at once,** *working perfectly together,* **or you're** *not* **alive. Period!!** Evolution is such a scientific JOKE!!

"<u>With</u> Oxygen in the air, the first amino acid would *never* **have gotten started, with<u>out</u> Oxygen, it would have been wiped out by cosmic rays."** - *Francis Hitching (Evolutionist - "The Neck of the Giraffe" 1982 p. 65)*

The situation would be *no different* **for the simplest of life forms. Everything would have to be in place,** *instantly.* **In structure, arrangement, coordination, coding, chemical makeup, feeding, elimination, respiration, circulation, and all the rest, - everything would have to be in perfect order - right at the start!**

All this would have *had* **to happen almost -** *instantaneously,* **or a couple minutes later the organism would die.** *Millions of functions* had to magically converge, all at the *same* time. *Another* **organism (of the** *opposite* **sex)** *also* **had to be created at about the same time and place, so they could mate and produce offspring.** Even worse for the shaky evolutionary theory is that every bit of this happened by complete *accident.* Yeah, right. - *EvolutionFacts.com*

Of course this *never* happened. As you can plainly see, it's utterly impossible. Yes, evolutionism is indeed a wild fairytale that sadly continues to lead a gullible world astray.

"The belief that life on Earth arose spontaneously from non-living matter, is simply a matter of *faith* **in strict reductionism and is based entirely on** *ideology."* **-** *Hubert P. Yockey (Non-creationist. Information Theory and Molecular Biology, Cambridge University Press, UK, p. 284, 1992)*

If You Believe in Evolution

In other words, you believe that every living thing was created from nothing, which somehow became a single cell. Thus all living things are related to each other... Yes - you being related to a banana, worm, dog, monkey, elephant, etc), BUT you do *not* believe it could start from an all powerful, *Intelligent Creator,* because that doesn't make sense to you...

LET ME ASK YOU THIS SIMPLE QUESTION...

How can you believe the ludicrous idea that *everything* **alive originated from absolutely** *NOTHING...* **but that an Intelligent Design** *Creator* **could** *not* **be possible?** Seriously? THINK about it for a while, a loooooong *evolution minute* if necessary.

Check out this quote in 1861, by evolutionist Adam Sedgwick about the reason why the theory of evolution was created. **"From first to last it is a dish of rank materialism cleverly cooked up... And why is this done? For no other reason, I am sure, except to make us independent of a Creator."** - *Adam Sedgwick (British geologist and Professor of Geology at Cambridge, England)*

Are You a Skeptic?

If you are a *skeptic of creationism...* you have been hoodwinked by one of the biggest lies of all time and you owe it to yourself and loved ones to *explore the truth about creation.*

In *recent* years, some evolutionists are *now* saying... *"Well of course dogs came from dogs, cats from cats, birds from birds, etc. - but there certainly was 'not' a Creator involved."* To which the creationists respond, *"So Mr. Evolutionist, how, where, and why did all those kinds of animals and plants begin then? And, where are the multiple-BILLIONS of transitional fossils that MUST be found to prove evolution is true?"* *"Why did you change your thinking on this when you were so sure it was correct all along?"* And, *"Did you know the Bible has stayed consistent about creation?"* Why? Well, for starters, because the Bible is the *truth,* and evolution is *not.*

Don't get me wrong, most creationists love the *people* who believe in evolution... they *don't* love the blatant (un-scientific) lies their insignificant theory exposes to mankind. **Creationists just can't stand by and allow humanity to be sold the evolutionary lies on how we all came to exist. It's time we all unite against the grand fairytale.**

Let's continue to learn why you can *intellectually* believe and trust that we were all *created* by a loving, all-knowing, intelligent Creator, with whom you can spend eternity. A *perfect* eternity, like the one talked about in the Bible, where everything is, well, PERFECT!! No more problems or worries. Can't believe it? Too much to imagine right now? That's cool, stick with me and you may change your mind by the end of this book. This isn't mumbo jumbo religious talk. *Billions* of believers around the world simply call this *intelligent design creator,* GOD. He's your creator who loves YOU very, very much and wants to spend *eternity* with you, and no matter what you have done in the past, He'll forgive you if you sincerely ask

Him to. He'll accept you just like you are right now. **You don't need to change first, He wants you just as you come to Him.**

IF THERE IS A PERFECT *ETERNAL* LIFE... waiting ahead for you and your loved ones... don't you think that would be **the *most* important thing in life that you need to be clear on?** NOW!! TODAY!! Of course it is. Nothing else in this life comes remotely close. That's why you need to finish reading this book, so you can learn what school *didn't* teach you. While you're at it, **start reading the Bible. It's *God's "love letter" to YOU* and the most important book you will ever read. *If God is real and eternity is perfect, then what could possibly be more important for us all than to know HIM, and commit our lives to HIM?***

One of the greatest scientists of the last 200 years, a Bible believing Creationist, said:

"Mathematics and dynamics fail us when we contemplate the Earth, fitted for life but lifeless, and try to imagine the commencement of life upon it. This certainly did *not* take place by any action of chemistry, or electricity, or crystalline grouping of molecules under the influence of force, or by *any* possible kind of fortuitous concourse of atmosphere. We must pause, face to face with the mystery and *miracle* of creation of living things."
- *Lord Kelvin (Quoted in - Battle for Creation, p. 232)*

Evolution takes a whole lot more *blind faith*... than believing in an *Intelligent Creator.*   Contrary to what some of you may think, I'm doing my best *not* to be judgmental or hypocritical about what I'm writing and citing in this book.

For the record, not only are Christians hypocritical and judgmental many times, so is *everybody* else. You are. I am. Even your mom is. We are not anywhere close to perfect or claim to be (unless we're crazy), but then neither is anybody else. **If you feel as**

if most Christians are hypocritical, and that's why you won't go to church or believe in God - get over yourself. It's a lame excuse not to go to church or to become a Christian. *Truth is truth and facts are facts.* We are *all* hypocrites. Jesus said He is the *only* perfect one, claiming to be both God in the flesh and the son of God. He said, **"I and the Father are one."** - *John 10:30*

That's why I suggest that **if you doubt intelligent design creation and are sold-out on evolution, ask for actual** *evidence* **about the** *evolutionary theories,* **and also look at** *intelligent creation* **closer. Study the Bible. Do your homework on both sides, not just one. Then you'll be able to make an** *intelligent* **and** *more informed* **life decision.** *The most important decision in life.*

Creationists *also* **believe in** *natural selection,* however, we know **it only** *"selects,"* **it does** *not* **add any** *new information* **which is required for evolution to be true.** More on this later as the stale, frail and pale theory is debunked even more.

"Evolution is a fairy tale for adults." - *Jean Rostand (Agnostic, and famous French biologist)*

Darwinism *(Definition)*

Darwinism is the *"unproven"* theory of the evolution of species by *natural selection* popularized by Charles Darwin. Followers of the sickly theory are often called *Darwinists* or *Darwinians.*

"Evolution is *unproved* **and** *unprovable.* **We believe it, only because the only alternative is special creation, and that is unthinkable."** - *Sir Arthur Keith (Evolutionist / Darwinist. Author of the foreword for the 100th edition of Darwin's book "On the Origin of Species")*

Neo-Darwinism *(Definition)*

Neo-Darwinism is the defective theory of the evolution of species by natural selection with the addition to and emphasis on *mutations,* which lead to evolution. We discuss in the "mutations section" of this book how this is utter nonsense and *cannot* possibly make life evolve into new species, with its necessary *new* genetic information (*mathematically* or *scientifically* speaking), in a trillion years.

The Third Way

If evolution were true, Darwinism and Neo-Darwinism would be the *only* two real appreciable choices. Since *Darwinian* and *Neo-Darwinian* evolution has been debunked over and over for over 160 years now, recently another version has popped up called, **The Third Way.** First of all, it's *not* actually a third way. It's just a *new twist* on natural *random accident* evolution. That's all. What's new? It's *repackaged Darwinism* and it's really *not fooling any thinking person,* so there's really no need to waste your time or mine on it in this book.

Three Main Kinds of Evolutionists

1) **DARWINISTS...** Darwinists **adhere to Darwin's idea that** *natural selection* **is the sole mechanism** through which evolution occurs. In a later book, Darwin rejected it and returned to *Lamarckism.*

Jean-Baptist Lamarck laid the foundation of modern evolutionary theory, called **LAMARCKISM,** which is the concept of *inheritance of acquired characteristics.* Many evolutionists still think this is a huge deal that supports their theory. An example is the giraffe, which he claimed got its long neck by stretching it up to reach the higher branches for its food.

To which I ask, are you DRUNK? This is beyond stupid. Why would it have to reach up high in the trees for its food? All of the other animals of the day who lived in the same area didn't have to reach high for food. And, the giraffe certainly didn't have to reach up for its *water*, because as we all know... *water doesn't grow on trees.*

He also believed in *UNIFORMITARIANISM...* which is basically **the *opposite* of what Creationists believe,** which is that the Earth had a ***worldwide flood*** *(which fits perfectly with the fossil record).* *Uniformitarianism* does *not* subscribe to the big flood. If it were true, we would find massive amounts of fossils forming *today,* but we *don't.* This theory (rather, an uneducated and unproven *guess*) states that the way that everything is occurring today is the way it has *always* occurred on our planet. Just another big, made up word, to try to impress people to believe in evolution. Whatever it takes I guess.

2) NEO-DARWINISTS... Neo-Darwinists declare that the mechanisms by which evolution occurred and are now occurring are ***mutations*, which are then refined by *natural selection.***

3) "HOPEFUL MONSTER" ADVOCATES (aka: *punctuated equilibrium* and *quantum speciation*)... Hopeful Monster Advocates pin their hopes on *sudden, massive mutations,* **producing a new species *all at once*** (Imagine a reptile laid an egg and a cute furry little baby raccoon hatched out of it, or a tiny little bird with wings and all). Their view is that a *billion-billion* beneficial mutations occur every 50,000 years in *two* newborns around the same time period - a male *and* a female of course - and they must somehow *magically* be located a short geographical distance apart so they just so happen to find each other *(fall in love and get married),* and then mate to propagate the so-called new species. This apparently happened for *each* of the millions of *different* species. Alrighty then. This theory is so whacked out, the fairytale goes even deeper down the rabbit hole. Really imaginative people come up with these wild and crazy ideas and others actually believe them, but then they refuse to believe in God. You can't make this stuff up. Oh wait, apparently you can. ;-)

"Evolution not only conveys *no* knowledge but it seems somehow to convey *Anti-Knowledge*." - *Colin Patterson (Evolutionist. Address at the American Museum of Natural History (Nov. 5, 1981).*

The Male / Female... *Requirement*

Inherent in the species quandary is the *male and female* element problem. It requires *both* a male and female to produce offspring.

"He who made them at the beginning 'made them male and female.'" - *Matthew 19:4 (The Bible)*

In 1984 it was supposed that mingling two sets of *genes* would produce a *new creature*. Researchers tried to fertilize *mouse eggs* with equal sets of *mouse genes* from other females. It *required* a *male* gene.

How could two of each species, independent of each other, evolve? Yet this is what *had* to happen. The male and female of each species are uniquely separate from one another in many ways, yet *perfectly matching partners*. **A *male and female* would have had to evolve *together*. Evolution can*not* explain this.** - *EvolutionFacts.com*

Almost all forms of complex life have both male and female; horses, dogs, humans, fish, moths, monkeys, elephants, birds, etc. **The male needs the female to reproduce, and the female needs the male to** reproduce. *One cannot carry on life without the other.* But if evolution were true, which came first, according to the dead theory?

If a male came into being before a female, how did the male of each species reproduce without females? How is it possible that a male and a female each spontaneously came into being, yet they have complex, complementary reproductive systems? **If each sex was able to reproduce without the other,** *why* **(and** *how***) would they have developed a** *reproductive system* **that** *requires* **both sexes in order for the species to** *survive?* - *Ray Comfort (Science Facts in the Bible)*

"He who thinks half-heartedly will *not* believe in God; but he who really thinks *has* to believe in God." - *Sir Isaac Newton*

Evolution Has ZERO Probability to Create Life *(NO Chance in a Zillion!!)*

Evolutionists talk about *probabilities* as if they were *possibilities*. They say that all they need is enough *time*. Dream all they want, but this is *not* true in the case of evolution.

TAKE THIS TO THE BANK... **How long would it take to walk across the universe? Well, after you had done it, you would need to do it billions of times more before you would even have time to try all the possible** *chance combinations* **of putting together just ONE,** *properly sequenced, left-only, AMINO ACID PROTEIN,* **in the** *right order.* **That's inside** *each* **cell, and there are estimated to be somewhere between 37** *trillion* **and 100** *trillion* **cells in your body.**

Thus, besides no *scientific chance,* **there is ZERO** *mathematical chance* **of life forming from non-life. The ONLY possible option is** *Intelligent Design* **(GOD).** (Note: I can't find the *citing* for this mathematical statistic. That shouldn't matter to evolutionists anyway, since their theories have *no evidence* and that doesn't seem to bother them one bit).

If you choose to believe in the lifeless theory of evolution after learning the real science and mathematics stated in this book, then so be it, but don't you think you owe it to yourself to at least be open to learn about the *facts,* instead of deficient fantasies?

"Unfortunately, in the field of evolution most explanations are *not* good. As a matter of fact, they hardly qualify as explanations at all; they are *suggestions, hunches, pipe dreams, hardly worthy of being called hypotheses."* - *Norman Macbeth (Evolutionist. Darwin Retried (1971), pg. 147)*

Natural Selection (Only "Selects")
It's the "Opposite" of Evolution

Natural Selection is another *huge WIN for Creationism* (*not* for evolution, as evolutionists would like to think and want you to believe). *Natural Selection* is actually a big *loser* for evolutionism.

You see, in reality ***natural selection*** (*reshuffling* or *losing* genetic information that's *ALREADY* there) **- is actually the *OPPOSITE* of evolution** (which *requires* NEW genetic information being *added* to the gene code, which we *never* see).

According to *evolutionary theory,* not only is *natural selection* said to have produced everything, but the entire process is said to be entirely RANDOM, every time! Therefore it is *not "selection,"* for *nothing* was *selected!* Just whatever happened next is what happened.

Random variations and *chance accidents* are said to have produced all the wonders around us. **The theory would be better called *"NATURAL RANDOMNESS,"* instead of *"Natural Selection."***

How can Total *Randomness*... select only that which is better, and move only in advantageous directions? Random occurrences *never* work that way. Yet in the never-never land of evolutionary theory, they are said to do so.

DARWINISM: THE BASIC TEACHING... When a plant or animal produces offspring, *variations* appear. Some of the offspring will be different from other offspring. Some evolutionists *(Darwinian evolutionists,* also called *"Darwinists")* declare that it is these variations (which they call *"natural selection"*) - *alone* - which have caused *all* life forms on our planet, such as: pine trees, jackals, clams, zebras, frogs, grass, horses and every other living thing.

Like evolutionists, *creationists also believe in natural selection,* however, as we mentioned already, we know **it only *selects,* it does *not* add, and NEW *genetic information,* which is REQUIRED *for evolution to be true. Evolution exposed* AGAIN** for *false reporting* and passing on *unscientific* information as if it's factual! Naughty!!

A Fundamental Teaching of Darwin's Evolution is... that *every living thing* in our world - whether it be a plant, animal, bird, or human... evolved from *other* creatures, which ultimately *originated* from *sand* and *water*... from a slimy mud pit. Come on, do *you* still really believe this preposterous fairytale? Please tell me, "NO!!"

Charles Darwin said that *natural selection* was the primary way that everything changed itself from lower life forms to produce new *(completely different)* species. Many evolutionists now *add* that *"mutations"* are a way of creating new species, as their evolution hypothesis *evolves* over time. It's pretty funny how ***Biblical***

Creationism always stays consistent, yet evolutionary theories *change (evolve)* at the drop of a hat, just to fit their ever-changing worldviews.

SUB-SPECIES... Changes occurring *within* a species, called *sub-species* (i.e. dogs *always* remaining *dogs*) are *not* evolution, but rather, *adaptation.* Evolutionists *try* their hardest to sell *modern day adaptation* to us as if that makes changes from one species into a completely new one *millions of years ago,* true. (such as: reptile to bird, fish to land animal, bear to whale, monkey to man, etc). However, this is complete *unscientific* nonsense, with *no proof* whatsoever. Don't buy the snake oil they're selling.
- EvolutionFacts.com

SURVIVAL OF THE FITTEST... has *nothing* to do with evolution, in fact, it actually achieves the exact *OPPOSITE* of evolution! It's simply about surviving, *not* evolving. Evolutionists couldn't be more *wrong* about their clueless theory on *survival of the fittest,* just like their good 'ol Uncle *Charlie Was Wrong.* Just sayin'

Evolutionary theory is... *one of the most dangerous, most insidious, theories ever unleashed upon mankind.* See it for what it really is, not for what it pretends to be.

"Anything that has evolved by *natural selection* should be *selfish.*"
- LIFE ("How Did it Get Here?" - 1985 - p.177)

The Peppered Moth
Cha Cha Cha Changes

Many evolutionists claim the changes from white moths to black moths in the mid 1800s in England, is the *best proof* for natural selection, thus evolution. Yikes, if that's the best so-called *evidence* they have, they might as well go home, 'cause it's... Game Over!!

Interestingly, evolutionists strictly maintain, as part of their creed, that the evolutionary process is *not* reversible. Part of this irreversibility idea *requires* that when one creature has evolved into another, the new creature *cannot* evolve back again!

This creates a *huge* problem for the *Peppered Moth* in England. You see, the peppered moth is by far the most frequently discussed evolutionary alleged "proof" of *natural selection,* so we'll spend a little more time with it.

Talking about the Peppered Moth, the evolutionary tilted *International Wildlife Encyclopedia (1970 edition), Vol. 20 says,* **"This is *the most striking evolutionary change* ever to have been witnessed by man."**

Noting that Darwin was plagued by his *inability* to demonstrate the evolution of even *one* species, Jastrow said: **"Had he known it, an example was at hand which would have provided him with the proof he needed. The case was an exceedingly rare one - the peppered moth."** - *Robert Jastrow, Evolutionist - Red Giants and White Dwarfs, p. 235.*

And, in his large 940-page book, ***Asimov's New Guide to Science,*** evolutionist *Isaac Asimov* mentions that some fools oppose evolution, saying it has never been proven; and then Asimov gives us a *single,* outstanding evidence: the *peppered moth.* (This is astounding - in view of the fact that it is *no* evidence at all! Isaac Asimov is the leading evolutionary science writer of the mid-twentieth century. If the peppered moth is the best he can come up with in defense of evolution, surely evolutionists have absolutely *no* real case, or clue, whatsoever).

NOW LET'S LOOK AT THE TRUTH... Before 1845, near Birmingham, England, the peppered moth was *primarily* light colored, but *some had darker wings.* (These darker varieties were called the melanic or carbonaria forms). **In accordance with Mendelian genetics, some peppered moth offspring were always born with light-colored wings while, others had darker wings. Thus it had been for** *centuries.* The little moths would alight on the light-colored tree trunks; and birds, able to see the darker ones more easily, ate them and tended to ignore the light-colored varieties. Yet *both varieties continued to be produced.* But then the industrial revolution came and the trees became darker from smoke and grime—and birds began eating the lighter ones. Note that peppered moths regularly produced *both* varieties as offspring.

By the 1880s in the Manchester, England area, toxic gases and soot were killing the light-colored lichen on the trees and darkened even more of the tree trunks. **The changeover from light to dark moths began there also.** The smoke and smog from the factories darkened the trunks of the trees where the moths rested. **This darkening of the trees made the dark-hued moths difficult to see and the lighter ones quite easy for the birds to spot.**

By the 1950s, 98% of peppered moths were the dark variety. *All the while, the moths continued to produce both dark "and" light varieties.*

Evolutionists point to this as a "proof of evolution," but it is NOT a proof of evolution by any means. We all know that there can be *variation* **with species.** *Variation within a species is NOT evolution.*

There are dozens of varieties of dogs, cats, and pigeons. But *no new species have ever been produced.* They are still dogs, cats, and pigeons. The same goes for *every* living organism (and *plant* as well).

There can be light peppered moths and dark peppered moths, but they are *all* still peppered moths. Even as **Asimov admitted, they are but variations within a *single* species.** The name of the single species that includes them both is Biston betularia. **They are ALL** *peppered moths,* **nothing more and nothing less.**

When (evolutionist) Harrison Matthews wrote the *introduction* for the 1971 edition of Charles Darwin's book, *The Origin of the Species*... *he DENIED the possibility of evolution in several respects,* and made this accurate observation about the peppered moth:

"The [peppered moth] experiments beautifully demonstrate natural selection - or survival of the fittest - in action, *but* they do *not* show evolution in progress, for however the populations may alter in their content of light, intermediate, or dark forms, *all the moths remain from beginning to end Biston betularia.*" - *Harrison Matthews ("Introduction," to Charles Darwin's Origin of the Species (1971 edition), p. xi)*

Let us consider this matter more closely: Because of dominant and recessive genes (Mendelian genetics), this little moth continued to produce both light and dark offspring for thousands of years while the birds kept eating the dark varieties... **Yet all that time, dark ones continued to be born! This is** *proof* **of the** *stability* **of the species... which is** *exactly the OPPOSITE of so-called evolutionary "proof!"*

Get it? Here again, for the umpteenth time, evolutionists say whatever they want, withOUT *real* **evidence of course (as usual), and expect a gullible world to gladly believe them and pass it along as** *evolutionary gospel.* **Yet they laugh at creationists.** Oh my, oh my. A strong case of the pot calling the kettle black. *And, then white, and then black again.* Give up already.

So, if you're falling for senseless lies like these, or you're a creationist, stop them in their tracks and demand *"real evidence."*

For nearly a century, the birds ate the lighter moths, but the darker ones kept being born. In recent years, industrial pollution laws are making the air cleaner, and the darker ones are more frequently eaten.

AGAIN... this is <u>NOT</u> EVOLUTION in the slightest, but simply a color change back and forth *within* a *stable* species. In fact, the pseudo famous photos taken back in the 1880s were exposed as fakes, They were *dead moths glued to tree trunks* for the photographs.

Two leading British evolutionist scientists said this about evolutionary claims for the peppered moth:

"We doubt, however, that anything more is involved in these cases than the selection of *already* existing genes." - *Fred Hoyle and *Chandra Wickramasinghe, Evolution from Space (1981), p.5.*

When he wrote his book, *Origin of the Species,* Charles Darwin gave many examples of variation *within* species, and tried to use them to prove evolution *outside* of true species. All this was before the discovery of *Mendelian genetics,* the *gene,* the *chromosome, DNA,* and the *DNA barrier* to evolution across basic types. In his ignorance Darwin wrote down his *imaginary* theory; and evolutionists today cling to this 160+ year old *hypothesis,* fearful to abandon it. - *EvolutionFacts.com* (Note: Colorful comments added by the author of this book, as in several other citings in this book).

Never, Ever Across Types

Plant scientists have bred unusual varieties of roses, corn, chrysanthemums, etc., but **NEVER do any of their experiments go *across* basic types.** As we study wildlife, we find the same thing: *NEVER* does one basic species change into another species.

There is *no* evidence that at any time, in all the history of the world, even one new true species has formed from other species. Yet evolutionary teachings require that such dramatic new changes would have *had* to occur thousands and thousands of times.

"Ultimately, the Darwinian theory of evolution is no more nor less than the great cosmogenic *myth* of the twentieth century." - *Michael Denton (Evolution: A Theory in Crisis 1985), p. 358)*

MUTATIONS *(Obliterate the Anemic Theory of Evolution)*

Evolution scientists have found the case for *natural selection* weaken so much over the years, that **now many have turned to *mutations* as their superhero *mechanism of change* and liberator from intelligent design creation.** Next we'll uncover *(using real science and math probabilities)* why **MUTATIONS ACTUALLY DISPROVE EVOLUTION,** thus destroying the listless theory of evolution even more.

If mutations produce only *negative* effects, and natural selection only *removes* negative effects - how can evolution result? It *cannot*.

Natural Selection and *Mutations* are the two most popular means by which Darwinian *macro*-evolution (one species evolving into a *different* kind) supposedly occurs. *Migration* and *Genetic Drift* are two other mechanisms of change that evolutionists like to talk about, but are *unworthy* and *unproven* to even be considered as possibilities, because they **only produce variations within the** *same* **species** *just as natural selection and mutations do.* (As discussed earlier in this book, this is called *"Adaptation."* Evolutionists call it *Micro-Evolution,* but it's *not* evolution at all. It's simply variations within the *same* species). Let's keep this topic to *mutations* for the time being.

Creationists believe in *adaptation,* **because we see it happening all over. Proof!! Evidence!! However, it does** *not* **in any way prove that one species changed/changes into** *another*. Of course actual *evidence* would support the harebrained theory of evolution (*one KIND evolving into another*; such as a fish into a land animal, or apple tree into a banana tree). However, evolutionists have *none*!! **"What good is half a jaw or half a wing? - These tales, in the 'Just-So Stories' tradition of evolutionary natural history, do** *not* **prove anything - concepts salvaged only by facile speculation do not appeal much to me."** - *Stephen Jay Gould (Evolutionist. "The Return of the Hopeful Monsters," Natural History, June/July, 1977)*

A MUTATION is *"damage"* **to a single DNA unit (a** *gene).* FAR from being beneficial, mutations constitute something *terrible* that *ruin and destroy organisms,* either in the first generation or soon after.

Not only is it *impossible* **for mutations to cause the evolutionary process... they actually** *weaken* **or** *terminate* **the life process! Mutations merely scramble** *existing* **genetic information, they** *NEVER* **create** *NEW* **information, which is** *NECESSARY* **for evolution to be true.** *This fact alone destroys the entire evil theory.*

Discovering that *so-called natural selection* accomplished no evolutionary changes whatsoever, modern evolutionists moved away from *Darwinism* into what they call, ***Neo-Darwinism*. This is their revised teaching that *Mutations PLUS Natural Selection* (not** natural selection *alone*) **produced all life forms on planet Earth.**

Mutations are rare, random, almost *never* an improvement, always weakening or harmful, and are **often *fatal* to the organism or its offspring.**

The problem here is that those organisms which mutations do not kill outright, are generally so weakened that they or their offspring tend to die out. Mutations, then, work the <u>opposite</u> of evolution. Given enough mutations, life on Earth would *not* be strengthened and helped; it would be *extinguished.*

<u>NO</u> EVIDENCE FOR MUTATIONS... "I have seen *NO evidence* whatsoever that these [evolutionary] changes can occur through the accumulation of gradual mutations." - *Lynn Margulis (Famous evolutionary biologist. Science Vol. 252, 19 April 1991, p. 379. Note: She has been awarded membership in the National Academy of Sciences, and is the **ex-wife of infamous Carl Sagan**)*

If scientists *really* believed in mutations as the great improvers of the race, they would ask that more, not less, mutagenic radiation might be given to plant and animal life! But **they well-know that mutations are *extremely dangerous*. Who is that confirmed Neo-Darwinist who is willing to let his own body be irradiated with X-rays, so that his offspring might wonderfully improve?** Nobody raising their hand? You think they'd be jumping at the chance. How about you?

Because of their random nature and negative effects, **mutations would destroy *all* life,** were it not for the fact that in nature they rarely occur.

What if... mutations were *plentiful* and always with *positive results,* but still as *random* as they now are? Guess what… **They would still be entirely *USELESS*.**

Even assuming random mutations (remember, they are *all* random) *could* produce those very complex structures called *feathers*... many birds would have wings on places like their stomachs, where they could not use them, or the wings would be upside down, without lightweight feathers, and under or oversized.

Most animals would have no eyes. Some would have one, some many, and those that had any eyes would have them in different places, like under their armpits or on the soles of their feet. I'll bet you never thought about things like this, have you?

The random effects of mutations would *annihilate* any value they might otherwise provide. - *EvolutionFacts.com*

H.J. Muller won a *Nobel Prize* for his work in *Genetics and Mutations.* Here is how he describes the problem:

"It is entirely in line with the accidental nature of mutations that extensive tests have agreed in showing the vast majority of them detrimental to the organism in its job of surviving and reproducing, just as changes accidentally introduced into any artificial mechanism are predominantly harmful to its useful operation. . . Good ones are *so rare* that *we can consider them all bad."* - H.J. Muller (*Nobel Prize* in *Genetics and Mutations*).

QUESTION FOR YOU… If you think mutations prove evolution, **what is it you know that this *Nobel Prize* winner didn't?** Keep in mind, you'll need a ton of actual *evidence* to back up your claim.

Mathematically Impossible

Fortunately mutations are rare. They normally occur on an average of perhaps once in every *ten million* duplications of a DNA molecule. **Even if** *all* **mutations were beneficial** (and again, almost all are NOT) **in order for evolution to begin to occur in even a very small way, it would be** *necessary* **to have,** *not just one…* **but a SERIES of** *closely related* **and** *interlocking* **mutations - all occurring at the** *same time* **in the** *same* **organism!** Yet another IMPOSSIBLE task to the highest scientific and mathematical degree!! (Read that again).

The odds of getting *two mutations…* **that are in some slight manner related to one another is** the product of two separate mutations: ten million times ten million, or **one hundred trillion to one.** What can two mutations accomplish anyway? Perhaps a honeybee with a wavy edge on a bent wing. But he is *still a little honeybee;* he has *not* changed from one species into another.

MORE *related* mutations would be *required.* The odds of three mutations in a sequence would be a billion trillion (1 with 21 zeros). But that would *not* begin to do what would be needed. Four mutations, that were simultaneously or sequentially related. But **all the Earth could not hold enough organisms to make that possibility come true.** And four mutations together does *not even begin* to produce real evolution. Millions upon millions of *harmonious, beneficial* characteristics would be needed to transform one species into another.

But ALL those simultaneous mutations would *have* **to be** *beneficial;* **whereas, in real life** (and worthy of repeating), *mutations very rarely occur and they are almost always harmful.*

OH, BY THE WAY… **you would also need to produce** *all* **those multi-mutations in MATED-PAIRS so they could properly**

produce young. Otherwise it would be like mating a donkey and a horse and getting *sterile* offspring.

Evolution *CANNOT* succeed *WITHOUT* **mutations, and evolution** *cannot* **succeed** *WITH* **them.** Thus, evolution is an *impossibility.* *Mutations* **are** *not* **your answer to life. And, for the record, life does** *not* **end when you die.** If you don't know him, *God is your answer* and He's waiting on you. (Like that random transition?). He loves you so much. Stop avoiding Him dear atheist and agnostic friends. He *won't* force you to love Him, He gives you *free-will* to choose Him *or* deny Him. I denied Him for the first 20 years of my life. Then one day many years ago, I asked Him to take control of my life. It's been an INCREDIBLE adventure ever since!!

"You will seek me and find me, when you seek me with all your heart." - *Jeremiah 29:13*

When Randomness Organizes Itself

Go to your local junkyard and ask that it be locked up and closed off for a year. Return from time to time to see if it cleans itself up and then arranges itself into an orderly collection of materials. Conclusion: *RANDOMNESS NEVER EVER ORGANIZES ITSELF.* How can it?

Incoherent matter in outer space could *never* **arrange itself into orbiting stars, galaxies, and planetary systems.** *EvolutionFacts.com*

You're thinking, *"Yeah, but that's not 'living' stuff."* No, but the point here is that *random does NOT become organized.* I revert back to… **show me the** *evidence.* Lots and lots of it while you're at it.

To paraphrase evolutionist Michael Denton… **"If complex computer programs *cannot* be changed by random mechanisms, then surely the same must apply to the genetic programs of living organisms."** - *Michael Denton ("Evolution: A Theory in Crisis" - 1985, p. 342)*

TIME... *(Is NOT Evolution's Little Friend)*

Many will say, *"Evolution is possible - if given enough time."* Evolutionists offer us almost 5 *Billion* years for mutations to do the job of producing all the wonders of nature that you see about you. But 5 billion years is, in seconds, only 1 with 17 zeros after it (1×10^{17}). And, the whole universe only contains 1×10^{80} *atomic particles.*

So there is *no possible way* that all the universe and all time past could produce such odds as would be needed for the monumental task. Julian Huxley, the leading spokesman for evolution of the mid-twentieth century, said - "It would take 103,000 *changes* to produce just one horse by evolution." That is 1 with 3,000 zeros after it, **almost 3 times more than all the atoms in the *universe!* Or,** categorically impossible! - *Julian Huxley (Evolution in Action, p. 46)*

Evolution would *require* millions of *beneficial* mutations all working closely together to produce delicate living systems full of *fine-tuned* structures, organs, hormones, and all the rest. And, **all those mutations would *have* to be *non-random* and *intelligently* planned!** In no other way could they accomplish the many necessary tasks.

But, leaving the fairyland of evolutionary theory, headed back to the real world, which only has rare, random, and harmful mutations, we must admit that mutations simply *cannot* do the job. No way, no how.

There is no other way that life forms could invent and reinvent themselves by means of that *mythical process* **called evolution.**

"It is no secret that evolutionists worship at the - *Shrine of Time.* **There is little difference between the evolutionist saying 'time did it,' and the Creationist saying 'God did it.' Time and chance is a two-headed deity.** **Much scientific effort has been expended in an**

attempt to show that eons of time are available for evolution."
- *Randy Wysong (The Creation-Evolution Controversy (1976), p. 137)*

TIME is *NOT* **some** *Magical Substance.* Just what is time, then? Time is merely a lot of past moments just like the present moment. Imagine yourself staring at a dirt pile or at some seawater, at a time when there was nothing alive in the world but you. (Not hard for ardent evolutionists to do). Continue carefully watching the pile or puddle for a thousand years and more. Would life appear in that dirt or seawater? It would *not* happen. Millions of years beyond that would be the same. Nothing would be particularly different. Just piled sand or sloshing seawater, and that is all there would be to it.
- *EvolutionFacts.com*

Evolution Requires
Increasing Complexity

The theorists have decreed that evolution, by its very nature, must move upward into ever-increasing *complexity,* better structural organization, and completeness. Indeed, this is a cardinal dictum of evolutionists. Again, **evolutionists maintain that evolution can** *only move upward* **toward more evolved life forms, and that it can** *never move backward* **into previously evolved life forms.**

But, **in reality,** *mutations,* **by their very nature, tear down, disorganize, crumble, confuse, and destroy.**

Check out what this ardent *evolutionist* has to say about mutations...
"No matter how numerous they may be, mutations do NOT produce *any* **kind of evolution."** - *Evolutionist, Pierre Paul Grasse (Evolution of Living Organisms (1977), p. 88)*

Evolution *Requires* <u>New</u> DNA Info.

In order for a new organism to be formed by evolutionary change, *new* **information banks must be emplaced.** It is something like using a more advanced computer program; a *card* of more complicated procedural instructions must be put into the central processing unit of that computer. **But, haphazard,** *random* **results of mutations** *never, ever* **provide new, structured information.** *- EvolutionFacts.com*

Evolution *Requires* <u>New</u> Organs

It is not enough for mutations to produce changes; they *must* produce new *organs* also! *Billions* **of mutational factors would be absolutely** *required* **for the invention of just** *one* **new organ of a new species**.

IN THE END… **evolutionists can** *make up anything* **they want and expect us to believe it.** And, they do. Maybe their goal is fame and fortune, or maybe they won't accept the fact that there is an *Intelligent Creator* - but they reject that because they want to be the ruler of their own destiny. Is that you perhaps? The problem is they have *no evidence* for evolution, and most of the leading evolutionists know that as fact. As amply proven, believing that mutations create new genetic code to produce new life forms is *unscientific* and only a *fairytale*. If you're not a creationist by now, please stop believing this Darwinism *Theory of Mutations,* evolution nonsense. - *EvolutionFacts.com*

"Evolution is *not* a fact. It *doesn't even qualify as a theory or hypothesis.* It is a metaphysical research program, and is *not* really testable science." - *Karl Popper (science philosopher)*

Evolution Needs to Occur in ONE Generation

Not even one major mutation, affecting a large number of organic factors, could accomplish the task of taking an organism across the species barrier. Hundreds of mutations - ALL *positive* ones - and all working together would be needed to produce a new species. *This can never, ever happen!!* The reason: The formation of even one new species would have to be done all at ONCE - in a *single* generation, *not* over millions of years! - *EvolutionFacts.com*

Q) WHAT DO YOU CALL IT WHEN CHEMICALS IN A PRIMORDIAL SOUP EVOLVE INTO DARWINIAN SCIENTISTS? A) SOUP to NUTS.

Gene Uniqueness *Forbids* Species Change

The very fact that each species is so *different* from the others, completely forbids the possibility that *random mutations* could change them into new species. There are millions of factors which make each species *different* from *all* the others. **The DNA code barrier that would have to be crossed is simply far too immense.**

"If life really depends on each gene being as unique as it appears to be, then it is *too* **unique to come into being by** *chance* **mutations."** - *Frank B. Salisbury (Evolutionist. "Natural Selection and the Complexity of the Gene," Nature, October 25, 1969, p. 342)*

Genetic Draft

Genetic Drift is frequently spoken of as another professed "evidence" of evolution, but **even confirmed evolutionists admit it proves** *nothing* **in regards to evolution.** Genetic drift are changes in small groups of sub-species that, over some time, have become separated from the rest of their species. Oddities in their DNA code factors became more prominent, yet **they** *all* **remained in the** *same* **species.**

Evolutionist Frank Rhodes in his book (Evolution, 1974, p. 75) explains that all "genetic drift" refers to is *changes in a "sub-species"* of a plant or animal (or in a "race," which is a sub-species among human beings). **Even Rhodes recognizes that** *genetic drift* **provides** *no* **evidence of change from one species to another. All the drift has been found to be within species and** *never* **across them.** - *EvolutionFacts.com*

No *Transitional* Species Found, EVER

"Evolution *requires* **intermediate forms between species, and paleontology [the study of fossils] does** *not* **provide them."** - *David Kitts (Evolutionist. "Paleontology and Evolutionary Theory" in Evolution, September 1974, p. 467)*

The speciation problem is a GAP problem. We find that there are *absolutely no (zero) transitional forms* to fill the *gaps*. You know: half dog/half bear, half bird/half dinosaur, and multiple millions of other examples that are *not* found in the fossil record, *anywhere*. Not for lack of trying and/or falsifying, mind you.

In desperation, *evolutionists* have come up with a feeble answer: **"The transitions were made so slowly that they left no remains behind."** - *Steven M. Stanley (Evolutionist. "Macroevolution and the Fossil Record" in Evolution, Vol. 36, No. 3, 1982, p. 460)*

WAIT A MINUTE!! How can that be? **The more** *slowly* **the transitions, the** *larger would be the number of transitional forms* **that would be in the fossil strata for posterity to examine!** And, none other than Charles Darwin himself agrees with us!

"When we descend to details, we can prove that *no* **species has changed [we cannot prove that a single species has changed]; nor can we prove that the supposed changes are beneficial, which is the groundwork of the theory."** - *Charles Darwin (in Francis Darwin (ed), Life and Letters of Charles Darwin Vol. 2 '87, p. 210)*

It Would Take a MILLION YEARS to Make a Species

How can there be multiple millions of animal and plant species, when the evolutionists tell us it takes a *million years* to make just *one* of them? The numbers don't even begin to add up. Ponder with me. They say, "It takes a million years to evolve a new species, ten million for a new genus, one hundred million for a class, a billion for a phylum." That's usually as far as our imagination goes.

There just is *not* enough time for all of those species changes to have occurred. Evolutionary dogma states that nothing was alive on Planet Earth over 2 billion years ago, and that all the evolving of life forms has occurred within that brief time span.

LET'S LOOK AT THE MATH... Two billion is only two thousand million. If it takes a million years to produce one species change, there would only be time for 2000 new species to be produced. *Not* millions upon millions of different species. An evolutionist might reply that more than one species was changing at the same time in various parts of the world, and this is how all our present millions of species could evolve into existence in 2 billion years.

But, that is a major (and incorrect) over-simplification. What about the theoretical stair-step pattern from the first single-celled creature that made itself out of sand and seawater to man? That single stair-step progression alone would **REQUIRE *hundreds of thousands of major changes!*** Yet only "millions of years" are provided for all the changes to come about? Somebody's flunked their math classes.

Billions **of** *transitional species* **would** *have* **to occur in order to climb the evolutionary stairs from amoeba to man.** Those transitional forms simply do *not* exist; they *never* have existed. There are only *gaps* between the species. But **the transitional forms would have *had* to be there in order for evolution to have occurred.** Say what you want, but it could *not* take place without them. Period!

They MUST be Seen <u>TODAY</u>... If the transitional changes occur that slowly, then **there must be *vast numbers* of *transitional species "living" TODAY,*** as well as **etched into the fossil record.** But they are *not* to be found. ANYWHERE!! They do *not* exist; and have *never* existed. End of the Evolutionary Fairytale.

Billions upon billions of transitional species *must* be engraved in the fossil rock *and* exist in nature today, if evolution is true. Yet we see *none* of this. **Over a hundred years of frantic searching by thousands of eager evolutionists has *not* produced even *one* real**

transitional form! The transitions *cannot* be found since they have *never* existed. - *EvolutionFacts.com*

GOD IS <u>NOT</u> DEAD!! EVOLUTION IS DEAD, Charlie Was WRONG!! Prove me wrong. Show me the *transitional* fossils.

IRREDUCIBLE COMPLEXITY
(Proves Evolution is a Fairytale)

Irreducible Complexity (definition) - is a single system composed of several well-matched, interacting parts that contribute to the basic function, wherein **the removal of any *one* of the fundamental parts causes the system to effectively cease functioning.**

"If it could be demonstrated that any complex organ existed which could *not* possibly have been formed by numerous, successive, slight modifications, *my theory would absolutely break down.*" - *Charles Darwin (Origin of Species)*

With this statement, **Charles Darwin provided a criterion by which his fairytale of evolution could be *falsified.*** The logic was simple: **since evolution is a gradual process in which slight modifications produce advantages for survival, it *canNOT* possibly produce complex structures in a short amount of time. That is *not* even remotely possible. It's a step-by-step process which may gradually build up and modify complex structures, but it *cannot* produce** them *suddenly as would be needed for survival.* **This is yet one more way that the pretentious, hopeless evolution theory crumbles to pieces. See ya later alligator.**

See, you either believe that God made every living thing completely formed instantly, or over a *very short period of time* (9 months in human mommy's tummy), or you believe <u>un</u>intelligent "chance" did it over *millions of years* by *random accident mutations.* The dilemma for evolutionists is that THE ACTUAL EVIDENCE *DISPROVES* EVOLUTION <u>EVERY</u> TIME!!

Mr. Darwin (you remember, the guy with *no science degree),* meet *Dr. Michael Behe...* biochemical researcher and professor at Lehigh University in Pennsylvania. Dr. Behe claims to have shown exactly what Mr. Darwin claimed would destroy the weak theory of evolution, through a concept he calls *Irreducible Complexity.*

In simple terms, this idea applies to any system of interacting parts in which **the removal of any one part** (OR *lacking* that part) *destroys* **the function of the *entire* system. An irreducibly complex system, then, *requires* each and virtually *every component to be in place BEFORE it can function properly.* Do you see the magnitude of this evolution buster?**

As a simple example of *irreducible complexity,* Dr. Behe presents the humble *mousetrap.* It contains only five interdependent, very simple parts which allow it to catch mice: the wooden platform, the spring, the hammer, the holding bar, and the catch.

Every one of these components is absolutely *essential* for the function of a simple little mousetrap. For instance, if you remove the catch, you *cannot* set the trap and it will *never* catch mice, no matter how long they may dance over the contraption. Remove the spring, and the hammer will flop *uselessly* back and forth - certainly not much of a threat to the little rodents. Of course, removal of the holding bar will ensure that the trap *never* catches anything because there will again be *no* way to arm the system.

If a mouse trap won't work if *any* one of its *five* parts are missing, do you really think the human body with *trillions* of parts, has a chance when it had to start from scratch and build over millions of years? Of course *not*. To think differently is the definition of *insanity* or *stupidity*. Or, you simply *choose not to believe in God* so you must disagree to keep your evolutionary worldview intact. No room for debate here. OPEN YOUR *INTELLIGENTLY DESIGNED* EYES and see the *blatant wickedness* and *stupidity* of the theories of evolution!!

NOTE WHAT THIS IMPLIES... Let's go back to the beginning of life, and the evolutionary teaching that every new species took *millions of years* to evolve. **An irreducibly complex system** (every living thing, and many non-living things) **could *not* have come about in a *gradual* manner. The system absolutely *requires all the components* to be added at the *same time, and* in the *right configuration, before* it works at all.**

Sure, you could argue that *an automobile takes hours to build before it works, it's not instant.* That's *not* the issue. The point is that once it's built, if it's missing even a battery wire, it won't move. **On the other hand, a "living" thing is far, FAR more complex** in that as all the parts were forming (from the first cell forward), if millions of things were not in place from the first few seconds to *minutes,* it would die.

Eventually, over eons, we'd have a *female* magically produced somehow before procreation could begin in the mom's belly, so that's not in the picture. That's another very complex issue all on its own.
We humans are *extremely complicated...* We have: brains to think, eyes to see, ears to hear, mouths to speak, drink and eat, and hearts to

pump oxygen and blood through our bodies of bones, fat, muscle, veins, arteries, central nervous system, etc. - just to keep us alive.

And, some of you think this all happened by some kind of aimless, purposeless *accident?* You choose to believe in this fancy *fable* rather than in *intelligent, supernatural design?* That takes a lot of imagination, I'll give you that.

We *need* most of these body parts *just to stay alive* more than a few minutes. Imagine the first basic life forms having blood but no arteries to push oxygen out to the body parts (which supposedly evolved over *long periods of time*), and veins to bring the blood back (without either knowing their intended future function). Or having only 2 inches of arteries started while the blood spills out all over the inside of your body, that somehow evolved before the complete arteries and veins. **They would have to be *fully-formed* from the *beginning,* as with virtually all of our body parts.** This is just basic common sense.

Exactly how did blood, veins and arteries form from scratch, and for that matter, how did they know they needed each other? How did they know how to make themselves function properly? Seriously!! They *didn't.* How did blood know it needed to work with oxygen? It *didn't.* How did we live as these were all supposedly *forming* over millions of years? The answer is obvious, we *didn't.* We *couldn't.*

And, let us not forget that to reproduce, we would need a *local* mate of the *opposite* sex that evolved this way as well. We could ask these types of questions about hundreds of body parts. These kind of questions should be *asked,* and *answered,* with *evidence.*

This is *not* complicated. If you disagree, use your God given brain and common sense to *think* about it.

Even only ONE simple *"protein"* within a human cell (and it's estimated that we have between 37 *trillion* and 100 *trillion* cells in our body), has the mathematical chance of 10 to the 50th power of forming by *accident* (i.e. evolution). In other words, it's *COMPLETELY IMPOSSIBLE* for us to have evolved into the super complex flesh and bone machines we are, even if we had *zillions of years* available. Yes, that's even longer than a googol. :-)

In fact, Dr. Behe asserts that the complicated biological structures in a cell exhibit the exact same *irreducible complexity* that we saw in the mousetrap example. In other words, they are *all-or-nothing.* Either everything is there and it *works,* or something is missing and it *doesn't work.* Such a system *cannot* be constructed in a gradual way, it *won't work* until *all* components are present. Darwinism has *no* mechanism for adding all the components at once.

Remember, Mr. Darwin's mechanism is one of *gradual mutations* leading to improved fitness and survival. A less-than-complete system of this nature simply will *not* function, and it certainly *won't* help the organism to survive. Common sense stuff, gang.

Indeed, having a *half-formed* and hence *non-functional* system would actually *hinder* survival and would be selected *against.* *Natural Selection only preserves* or *"selects"* those structures which are *already* functional. It does *not* create. If it is *not* functional, it *cannot* be naturally *selected.*

THINK FOR YOURSELF!! *Evolution is the stupidest fairytale to ever come out of people's mouths.* Stop believing this insanity.

- A mashup of an Ideacenter.org article and this book's author.

Syntropy

In order for a creature to live, eat, survive, and reproduce, it must basically be fairly *perfect* so to speak. It *cannot* have only *part* of its structure, but *must* have *all* of it. And, that structure must be fully complete. **Of the millions of DNA codes within its cells, essentially** *all* **must be there in perfect lettering and sequence in order for it to live and function properly.** This *coding requirement* is called *syntropy,* and it stands as yet another impossible barrier to evolution across basic species.

Natural selection within a species may work fine, but *you have to have the traits to begin with!* These **traits may adapt (adapting traits to new situations is** *not* **evolution)... but the traits had to be there to start with.**

"Evolution *cannot* **be described as a process of adaptation because all organisms are** *already* **adapted - Adaptation leads to natural selection, natural selection does** *not* **necessarily lead to greater adaptation."** - *Lewontin ("Adaptation," in Scientific American, September 1978).*

Although it occurs all the time within species, *natural selection* does *not* explain the origin of species or traits, but *only* their *preservation* and *more careful use.*

Lewontin is a confirmed evolutionist, but he recognizes that natural selection could *not* possibly produce evolution:

"Natural selection operates essentially to enable the organisms to maintain their state of adaptation *rather* **than to improve it. Natural selection over the long run does** *not* **seem to improve a species chances of survival, but simply enables it to track, or keep up with, the constantly changing environment."** - *Ibid*

You *cannot select* what is *not* there. If the trait is not *already* in the genes it *cannot* be selected for use or adaptation. **Selecting which trait**

will be used (which is *natural selection*) is *not* evolution; for the trait was *already* at hand. - *EvolutionFacts.com*

MOM and DAD on... EVOLUTION VS. CREATIONISM

A little girl asked her mom, "How did the human race appear?"

Mom answered, "God made Adam and Eve and they had children, and so was all mankind made."

The next day the girl asked her dad the same question.

Dad answered, "Millions of years ago there were monkeys from which the human race evolved."

The confused girl returned to her mom and said, "Mommy, how is it possible that you told me that we were all created by God, and Daddy said we evolved from monkeys?"

Mom answered, "Well honey, it's simple. I told you about my side of the family, and your daddy told you about his."

HAHA

"For I know the plans I have for you," declares the Lord, "plans to prosper you and not to harm you, plans to give you hope and a future."

- Jeremiah 29:11

THE FOSSIL RECORD
(Completely Annihilates the Misleading Evolution Theory)

One of the strongest evidences *against* evolution for evolutionists to prove that their theory is true, is the *Fossil Record* (or *lack thereof*) which when examined, unquestionably *destroys* the deceiving theory of evolution!!

The actual fossil record is the exact *"opposite"* of the theory of evolution.

Think about it, since so-called evolution is not seen happening now, all that the evolutionists have to prove their theory, is any *fossil evidence* of life forms which lived in the *past*.

"I think, however, that we must go further than this and admit that the *only acceptable explanation* is *Creation*. I know that this is anathema to physicists, as indeed it is to me, but we must not reject a theory that we do not like if the experimental evidence supports it." - *H. Lipson (Evolutionist. "A Physicist Looks at Evolution," Physics Bulletin 31 (1980), p. 138)*

Fossils **are of the highest importance** to evolutionary theory, for they provide the *only record (proof)* of plants and animals in ancient times. In the fossil record, scientists should be able to find all the evidence needed to prove that one species has evolved out of another. They have *not*. - *EvolutionFacts.com*

Fossils *(Definition)*

Fossils are the remains of living organisms, both plants and animals, or their tracks. These are found in *sedimentary rock.* Sedimentary rock is composed of strata, which are layers of stone piled up like a layer cake. (FYI, strata is the plural of stratum). Sedimentary rock is fossil-bearing or fossiliferous rock. Let's *dig deeper* into this. (Yes, pun intended).

These fossil remains may be *shells, teeth, bones,* or *entire skeletons.* A fossil may also be a *footprint, bird track,* or *tail mark* of a passing lizard. It can even include *rain drops.* **Many fossils no longer contain their *original* material, but are composed of mineral deposits that have infiltrated them and taken on their shapes.**

One thing to keep in mind is that... **the *"extinction* of a species" is *not* evolution,** as is sometimes stated by desperate evolutionists!

I wish I could be a Dinosaur Fossil.
Only then would someone dig me.

Transitional Fossils *(Definition)*

Transitional fossils, also called *intermediate fossils,* are (imaginary) remains of creatures which *must* be found *in-between* one *kind* of creature and another (different) *kind,* such as in-between a reptile and a bird for example. An example would be a half reptile / half bird.

There are *no credible transitional fossils anywhere in existence today.* Sure, loads of *fakes,* but no real ones.

"If I knew of any evolutionary transitionals, fossil or living, I would certainly have included them in my book, *'Evolution.'"*
- Dr. Colin Patterson (Evolutionist and senior Paleontologist at the British Museum of Natural History, which houses 60 million fossils)

For Evolution to be True...

For evolution to be true... *every living thing must have ultimately evolved from other kinds of living things,* **going back to** *nothing* **as the beginning of it all.** *The fossil record shows ZERO evidence for this.*

Some will say there are thousands of *transitional fossils* in captivity, however, in reality **there are** *no* **transitional *(half rat, half bird, etc)* fossils in existence.**

Think hard about that for a minute… even with the 100 million to 200 million fossils on display in *museums around the world,* not counting the estimated trillions in the worldwide strata, **there would absolutely** *have* **to be** *billions* **of** *transitional fossils* <u>*found*</u> **for evolution to be true. All we find are** *FULLY-FORMED* **animals** *and* **plants from the lowest stratum to the very top.**

On the other hand… *if intelligent creation is true,* **all the different kinds of animals and plants should appear** *fully-formed* **and** *abruptly…* **as they all indeed do in the worldwide fossil record,** in all of recorded history, and still do today. We're not talking about a small number of fossils, but rather tens of millions in the museums, and hundreds of billions to trillions within the strata.

BOOM EVOLUTION… "Creation and evolution, between them, exhaust the possible explanations for the origin of living things. Organisms either appeared on the Earth *fully-developed* **or** *they*

did not... **If they did appear in a fully developed state, they** *must* **have been** *created* **by some** *omnipotent intelligence."*
- D.J. Futuyma (Evolutionist. Science on Trial (1983), p. 197)

NO EVOLUTION TODAY... Evolutionists admit that evolution *(one kind of animal changing into another kind)* never occurs today, nor has ever occurred in the past 100,000 years (or so say the evolutionists).

The Two Primary Viewpoints... A basic postulate of evolution is the concept of *Uniformitarianism.* According to this baseless theory, the way everything is occurring today is the way it has always occurred on our planet. This point has strong bearing on the rock strata. Since no more than an inch or so of sediment is presently being laid down each year in most non-alluvial areas (Alluvial means: clay, silt, sand, gravel, or similar detrital material deposited by *running* water), no more than this amount could have been deposited yearly in those places in the past. Since there are thick sections of rock containing fossils, those rocks and their contents must have required *millions of years* to be laid down. That's how the theory goes anyway.

The *opposite* **viewpoint is known as** *Catastrophism,* **and** teaches that there has been a *great catastrophe* in the past - a *Worldwide Flood,* which within a few months laid down all the sedimentary rock strata, entombing the animals contained within them, which became *fossils.*

The Fossil Record *alone* **unconditionally** *destroys* **the presumptive theory of Darwinian evolution** which says that basically *nothing* created *everything,* starting from a mud-pit over 4 billion years ago. In fact, we could end the book after this section alone and there is *no* possible way the theory (more accurately stated: the *hypothesis, the guesses, the fairytale stories*) of evolution could survive the blow.

"Above all, you must understand that in the last days scoffers will come, scoffing and following their own evil desires. They will say,

'Where is this 'coming' he promised? Ever since our ancestors died, everything goes on as it has since the beginning of creation.' But they deliberately forget that long ago by God's word the heavens came into being and the Earth was formed out of water and by water. By these waters also the world of that time was deluged and destroyed." - *2 Peter 3:3-6*

Fossiliferous Strata means *fossil-bearing* strata. *Only sedimentary rocks have fossils,* for they are composed of sediments laid down at the time of the great flood, which hardened under pressure and dried into rock. **You will *not* find fossils in granite, basalt, etc.**

As mentioned earlier (a few times, lol), most people who buy into the imaginary evolution theory, state that every living thing evolved initially from *nothing*. A Big Bang supposedly somehow *magically* happened 13.8 billion years ago, thus eventually evolving into *every* living thing that exists today.

They don't take into account how evolution is completely and utterly *impossible* because of these things: *Irreducible Complexity*, the *lack of ANY Transitional Fossils found anywhere on Earth, the Cambrian Explosion* and hundreds of other very powerful scientific facts. **Over a hundred years of frantic searching by thousands of evolutionary scientists and amateur bone diggers has *not* produced even "one" (real)** *transitional (intermediate) fossil.*

Those supposedly omniscient scientists who still teach evolution as though it were fact, are finally seen for what they are… *frail men and women willing to believe a lie because it helps them avoid the truth.*

"Faith is the substance of fossils hoped for, the evidence of links unseen." HaHa - *Dr. Scott Huse (The Collapse of Evolution)*

Fossils are extremely important to the ill-advised evolutionary theory, for they provide our *only* record of plants and animals in

ancient times. The fossil record is of the highest importance as a *proof* for evolution. In these fossils, scientists must be able to find all the *evidence* needed to prove that one species has evolved out of another. They *never have* and *never will.* - *EvolutionFacts.com*

The more you study the paleontological record, the more certain you become that evolution is based on *faith* alone. - Evolution is only *faith, a religion.* Don't make it out to be anything more.

Fossil Dating

No, I'm *not* talking about when a 25 year old woman is dating a wealthy 97 year old man. That's just wrong on so many levels.

"When dinosaur bones containing carbon are found, they are NOT '*Carbon Dated'* because the result would only be a few *thousand* years old." - *Dr. Kent Hovind*

First of all, *no* **dating method can accurately date something to be millions or billions of years old. There is** *no* **possible way to accomplish this.** *Don't* **accept a different story. It's all just a** *guessing game.*

Evolutionists try to prove long ages of time by certain theoretical dating methods. Yet as we analyze those dating methods, we find **each of them to be** *highly flawed* **and** *extremely unreliable.* Not even uranium dating can be confirmed. Some methods are *many millions of years* or more *apart* in their dating.

Apart from recorded history, which goes back no further than several thousand years, we have no way of verifying the supposed accuracy of theoretical dating methods. In fact, not even the dating methods confirm the dating methods! They all give *different* dates!

With but very rare exception, they *always* **disagree with one another!**

LONG AGES NEEDED... For nearly two centuries, evolutionists have known that **since there was *no* proof that evolution had occurred in the past and there is** *no* **evidence of it occurring today, they would need to postulate very** *long ages* **as the means by which it** *somehow* **happened!**

NO ANCIENT FOSSILS... One worthwhile discovery that scientists made when they applied *amino acid* dating methods (both amino acid decomposition and amino acid racemization) out in the field - was that *traces of amino acid still exist all through the fossil strata!* **This means that** *"none"* **of the fossils are ancient!**

"Shells from LIVING snails were Carbon Dated as being 27,000 years old." - *Science magazine (Vol. 224, 1984. Emphasis added as in several of the quotes in this book)*

Here's an example of a **2.8 million year discrepancy** *fossil dating blunder...*

"Even the lava dome of Mount St. Helens [produced in 1980] has been radiometrically dated at 2.8 million years." - *H.M. Morris (Radiometric Dating, Back to Genesis, 1997]*

Not Even ONE *Missing-Link* Found

Over a one-year period from 1980 to 1981, *Luther Sunderland* **interviewed three of the** *most important paleontologists in the world,* **who were in charge of at least 50% of the major fossil collections on the planet** (*That's between 100 million and 200 million fossils*), covering every basic fossil discovery in the past 150+ years.

He found that **not even "one" of them could name a single so-called Missing Link** (*a halfway species between our regular species*). (L.D. Sunderland, Darwin's Enigma, p. 89) - *EvolutionFacts.com*

NOTE: These 3 top paleontologists are *evolutionists* at the top of their field. **If you disagree with the statement above, what do you know that these top experts don't?**

The Cambrian Explosion
(SUDDEN APPEARANCE of Fossils Flat Out Kills Evolutionism)

"Why then is *not* every geological formation and every stratum full of such *intermediate* links? Geology assuredly does *not* reveal any such finely graduated organic chain; and this, perhaps, is the most obvious and serious objection which can be urged *against* the theory." - *Charles Darwin (The Origin of Species) (Italics added)*

Organisms either appeared on the Earth *fully-formed*, or they did *not.* If they did appear fully-formed (as all *evidence* shows in the *Cambrian* layer, the *lowest* (bottom/oldest) of the geologic column), then the *only* answer is they were put here by an *Omnipotent Intelligence.* In other words, GOD.

With *no evidence whatsoever,* evolutionists claim the Cambrian layer formed around 500 *million* years ago (*Give or take a few days. Just kidding, but that's how ridiculous they are with these timelines*), and expect, or at least hope, that everyone will blindly believe it.

WHY SUCH VERY <u>COMPLEX</u> FOSSILS AT THE BOTTOM? The Cambrian has invertebrate (non-backbone) animals, such as multiple billions of *trilobites* and *brachiopods*. These are both *extremely complex* little creatures. ***Many* of our *modern* day *animals and plants* are in this lowest level, too.** The funny thing is that **they look the same *then*, as they do *today*.** Huh? This fits perfectly with the creation story, but *not* so well for the evolution fairytale.

How could such complex, multi-celled creatures be there in the bottom of the Cambrian stratum? But there they are. And, *no transitional fossils lead up to them, or are found above them.* Huh? This *alone* tells us that evolution *never* happened! Period.

Suddenly, in the very lowest fossil stratum, we find very *complex* plants and animals - lots and lots and *lots* of them, with *no* evidence whatsoever found *anywhere* that they evolved from anything lower or earlier.

BOOM EVOLUTION... **"It remains true, as every paleontologist knows, that most new species, genera and families, and that nearly all categories above the level of families, appear in the [fossil] record *suddenly* and are *not* led up to by known, gradual, completely continuous *transitional* sequences."** - *George G. Simpson (The Major Features of Evolution, p. 360)*

Paleontologists (Fossil Hunters) call this immense problem *the "Cambrian Explosion"* ... **because vast numbers *(as in billions)* of complex creatures suddenly appear in the fossil strata from the lowest level (Cambrian stratum) to the top - with *no evidence that they evolved from any less complicated creatures!*** NONE! NADA! ZERO! ZILCH!

Here's a discovery that alone *completely defames* the anemic little theory of evolution. Darwin must be turning over in his grave.

Here's Yet One More EVOLUTIONISM SPOILER... In 1968, *William J Meister, Sr,* a non-Christian evolutionist who made a hobby of finding trilobite fossils found a *human footprint and trilobites* **in the** *same* **rock in the supposed** *500 million* **year old Cambrian stratum** (the lowest and oldest), in Utah. The trilobites were *right in the footprint.* The amazing thing is that **the human was wearing a "sandal."** It was later confirmed when *scientists found more sandaled footprint fossils.* Keep in mind, that according to the half-baked evolutionary theory, man is only about *200 thousand* years old. *(Creationists believe humans are about 6,000 years old).* This is *not* an isolated incident either. Boom, evolution just lost a few more followers to creationism. Go God!!

The Cambrian Explosion (AKA: "Sudden Appearance") proves **once and for all that evolution *NEVER* happened! Period! End of the Dumb Evolutionary Fairytale.**

Thousands of other real stories that debunk evolutionary theories are easily found if you look for them. Many can be found at CharlieWasWrong.com and several other sources listed earlier in this book.

ABRUPT APPEARANCE... The *smaller, slower moving creatures appear suddenly* at the *bottom* in the Cambrian. **Above the Cambrian, the larger, faster creatures appear just as suddenly!** And, when these life forms do appear, they appear by the *millions!* Tigers, salmon, lions, pine trees, gophers, hawks, squirrels, horses, and on and on. Evolution cannot explain this sudden emergence, and competent scientists acknowledge this fact.

"The abrupt appearance of higher taxa in the fossil record has been a perennial puzzle. Not only do characteristic and distinctive remains of phyla appear *suddenly, without known ancestors,* **but several classes of a phylum, orders of a class, and so on, commonly**

appear at approximately the same time, without known intermediates." - *James W. Valentine and Cathryn A. Campbell ("Genetic Regulation and the Fossil Record," in American Scientist, November-December, 1975)*

All evolution has is faith to go on, for there are no facts...

"The hypothesis that life has developed from inorganic matter is, at present, still an article of *faith."* - *J.W.N. Sullivan (The Limitations of Science (1933), p. 95)*

"The sudden emergence of major adaptive types as seen in the abrupt appearance in the fossil record of families and orders, continued to give trouble. The phenomenon lay in the genetic no-man's land beyond the limits of experimentation. A few paleontologists even today cling to the idea that these gaps will be closed by further collecting . . but most regard the observed discontinuities as real and have sought an explanation." - *D. Dwight Davis ("Comparative Anatomy and the Evolution of Vertebrates," in Genetics, Paleontology, and Evolution (1949), p. 74)*

"ABRUPT APPEARANCE" *PROVES* once and for all that evolution categorically *NEVER, EVER* happened! End of Fairytale!!

Simply Stasis

Stasis means that fossils in the fossil record stay the *SAME...* throughout history. Meaning, we don't see any "transitional" fossils anywhere in the fossil record, from the very bottom to the tippy top.

Throughout the fossil record we find the *very same creatures* that we have *today,* plus some extinct types which died out before our time!

As already discussed and worthy of repeating again and again, neither now nor earlier are there transitional forms, halfway between different species, found.

In the rock strata, we find lots of horses, tigers, fish, insects, etc. but *no* transitional forms. For example, we find large horses and small horses, but *nothing* that is part horse and part anything else. Seems quite odd if evolution is true, doesn't it?

"STASIS" also *PROVES* once and for all that evolution *NEVER* happened! Period! End of the Fake News Fairytale!!

After giving years to a careful examination of the fossil record, comparing it with that of species alive today, a famous biologist on the staff of the Smithsonian Institute wrote these words:

"All the major groups of animals have maintained the same relationship to each other from the very first (from the very lowest level of the geologic column). Crustaceans have *always* been crustaceans, echinoderms have *always* been echinoderms, and mollusks have *always* been mollusks. There is *not* the slightest evidence which supports *any* other viewpoint." - *A.H. Clark (The New Evolution: Zoogenesis (1930), p. 114)*

Most evolutionary scientists today have given up looking for *transitional* fossils, after searching doggedly for the last 160+ years. Why? Because they know they are *not* there.

If there are *no* transitional fossils in the fossil record (And we are now 100% positive that there are *none*), **then there has been NO evolution,** whether they want to admit it or not.

<u>UN</u>-**CHANGING SPECIES...** Going a bit deeper - *Stasis* **means to retain a certain form, to remain** *unchanged;* **in other words,** *not* **to change from one species to another!** The problem for the evolutionists is the fact that the animals in the fossil record did *not* change. **Each creature first appears in the record with a certain shape and structure. It then continues on unchanged for** *millions of years (according to evolutionary timeline),* **and is either identical to creatures existing** *now,* **or becomes extinct and disappears.** But all the while that it lived, there was *no* change in it; *no* evolution.

There is *no* **evidence of what paleontologists call** *gradualism,* **that is, gradual changes from one species to another.**

There was and is *only* **stasis.** The *Gap* *problem (no transitional forms between species)* **and the** *Stasis* *problem (species do not change)* *simply* **ruin the alleged theory of evolution.**

The history of most fossil species includes two features particularly inconsistent with gradualism:

STASIS: "Most species exhibit *no* **directional change during their tenure on Earth. They appear in the fossil record looking much the same as when they disappear; morphological change is usually limited and directionless."** - *Steven Jay Gould (Evolutionist. "Evolution's Erratic Pace," in Natural History, May 1977, p. 14)*

SUDDEN APPEARANCE: "In any local area, a species does *not* **arise gradually by the steady transformation of its ancestors; it appears** *all at once* **and** *'fully formed.'"* - *Steven Jay Gould (Evolutionist. "Evolution's Erratic Pace," in Natural History, May 1977, p. 14)*

Recapping Some of the Problems

Here are some of the key problems with the fossils in the strata. **These problems are serious enough that any *one* of them alone is enough to overthrow the unrepentant evolutionary theory in regard to paleontology and stratigraphy:**

(1) Life *suddenly appears* in the bottom fossil-strata level, the Cambrian, with *NO PRECURSORS*. THIS IS A HUGE INSURMOUNTABLE PROBLEM for evolution!! **This one proof alone *devastates* the theory.**

(2) When these lowest life forms appear; they are small slow-moving, extremely abundant, numbered in the *billions* of specimens, and quite complex).

(3) *NO transitional* species are found at the bottom of the strata, the Cambrian.

(4) Just below the Cambrian, in the Precambrian, there are *no* fossil specimens, thus no transitionals.

(5) *NO transitional* species are found above the bottom stratum, from the Ordovician on up to the top.

(6) Higher taxa *(forms of life)* appear just as suddenly in the strata farther up. These higher types (such as beavers, giraffes, etc.) *suddenly appear* with *NO hint of transitional life forms leading up to them.*

(7) When they appear, *vast numbers* are to be found.

"I myself am convinced that the theory of evolution, especially the extent to which it's been applied, will be one of the great *jokes* in

the history books of the future. Posterity will marvel that so very flimsy and dubious an hypothesis could be accepted with the incredible credulity that it has." - *Malcolm Muggeridge (well-known British journalist and philosopher)*

Here's Why the Fossil Strata Theory is a... *Colossal Hoax*

IMMENSE NUMBER OF FOSSILS... One of the most startling facts about the sedimentary strata around the world is the *vast quantities of fossils* they contain. **Without a Worldwide Flood,** it would be *impossible* for such huge amounts of plants and animals to have been rapidly buried. And, *without Rapid Burial* they could *NOT* have fossilized.

Sedimentary strata **are filled with fish fossils, yet when a fish dies** *today,* **it** *never* **fossilizes.** It bloats, floats, and then is eaten by scavengers and other small creatures. There have even been *shark fossils* found that are only ¼ **"inch"** thick. **ONLY a flood catastrophe such as Noah's Worldwide Flood, could do this.**

The strata have lots of animals in them; but, when an animal dies *today,* **it usually** *never* **fossilizes.** It *rots* if the buzzards do not find it first. **The multiple billions found in the fossil record** *scream* **"FLOOD!!"** There is *no other choice.* Think about this. Really, THINK about it for a second. It's just *Common Sense* my friends. Not a big mystery.

"Despite the bright promise that paleontology provides a means of 'seeing' evolution, it has presented some nasty difficulties for evolutionists, the most notorious of which is the presence of 'gaps'

in the fossil record. *Evolution requires intermediate forms* between species and *paleontology does not provide them..."* - *David B. Kitts (Ph.D. in Zoology, Head Curator, Department of Geology, Stoval Museum, and well-known evolutionary paleontologist. Evolution, Vol. 28, Sept. 1974)*

There is a huge abundance of fossilized *PLANT* life in the strata, too, yet when a weed, bush, or tree dies, it turns back to soil. It does *not* harden into a fossil.

It requires some very special conditions to produce fossils, which occurred *one* time in history. The evidence is clear that it was a worldwide phenomenon, and that it happened very, very quickly.

RAPID BURIAL... A striking fact about the fossils is that they were obviously all laid down at the same time - and very rapidly!

The fact is that fossils *never* form at the present time... yet, in the sedimentary strata, we find literally *multiple BILLIONS* of them! Examination of the strata bearing them reveals it was obviously laid down by a *MASSIVE FLOOD OF WATER.*

What happened? A *terrible catastrophe* occurred that suddenly overwhelmed the *entire Earth!* Large amounts of *fossil seashells* **have been found on and in the *highest mountains* of the planet,** including the highest range of them all, the Himalayas.

FISH SWALLOWING FISH... Princeton University scientists were working in Fossil Lake, Wyoming, when **they found a fossilized fish that was in the process of *swallowing another fish.*** Because *both fish had been pressed flat by the sudden burial*, the paleontologists could see one fish *inside* the other with only the tail sticking out of the larger one's throat.

Obviously, *this required a very sudden event* to capture and kill a fish swallowing a fish! Nothing like this happens *today*.

In the Hall of Paleontology, at Kansas State University, you'll find a 14-foot fish that swallowed a 6-foot fish. The fish that was swallowed was *not* digested, and then *both had been suddenly entombed.*

NO EVOLUTION TODAY... Evolutionists are forced to admit that Darwinian Evolution *(one type of animal changing into another, such as ape to man) never* occurs today.

"No biologist has actually seen the origin by evolution of a major group of organisms." - *G. Ledyard Stebbins (Geneticist. Process of Organic Evolution, p. 1)*

EVERYTHING HINGES ON FOSSILS... Clearly, then, because obviously *no* evolution is occurring in modern times, all that the evolutionists have to prove their erratic and offensive theory is fossil evidence of life forms which lived in the *past.*

If evolution is the cause of all life on Earth, then there *absolutely must be multiple billions* of various, partly evolved fossil life forms. For evolution to occur, this *had* to occur in *great abundance*. The fossils must reveal incredibly large numbers of transmuted species, creatures which are half-fish/half-animal, etc. Yet they *fail* to do so.

"Evolution... is not only under attack by fundamentalist Christians, but is also being questioned by *reputable scientists*. Among paleontologists, scientists who study the fossil record, there is growing *dissent* from the prevailing view of Darwinism." - *James Gorman (Evolutionist. "The Tortoise or the Hare?" Discover, October 1980, p. 88)*

POLYSTRATE (VERTICAL) **TREES...** Think about it a minute: **Could a *"vertical" tree* die and *stand* there for half a million years while rock strata gradually covered it? Of course it could *not*. Yet we find *polystrate trees* like this in the strata and even in coal beds.**

Upright tree trunks (polystrate trees), 10 to 30 feet or more in height, are *often found* in the strata associated with coal or in the coal itself. **The sediments forming the coal *had* to form *rapidly* in order to solidify before the tree trunks could rot and fall over. Sometimes these upright trees are even found *UPSIDE DOWN* in the strata, giving us all just one more reason to deny evolution.** Coal and oil can actually be made in a very short period of time, *unlike* the millions of years that evolutionists tell us. The key is *immense pressure,* such as what you would get from a *catastrophic flood.* By the way, these things do *not* happen today.

"Ah, Lord God! Behold, You have made the heavens and the Earth by Your great power and outstretched arm. There is nothing too hard for You." *- Jer. 32:17*

HELP US SPREAD THE WORD
JOIN US DAILY ON **FACEBOOK**
Please LIKE, COMMENT and SHARE

facebook.com/CharlieWasWrong

All *Intermediate* (Transitional) Life Forms Would be... *FATAL*

What good is *"half a wing," "half a brain," "half gills/half lungs,"* or *"half a heart?"* If reptiles evolved into birds, *where* are all the *transitional* fossils? Again, as with all kinds, there are *none to be* found. Only *fully-formed* reptiles and birds. Who would the first bird *mate* with anyway? They would die before evolving.

Evolutionists **tell us that the** *wing* **evolved four** *separate* **times: in insects, flying reptiles, birds, and bats. So why believe it just happened** *one* **time by a supernatural act of God, when you can believe it happened** *four different times,* **each being an** *unplanned, random accident?* **Why indeed. Please tell me that** *you* **don't buy into this little "theory."**

NATURE *Requires* an Explanation. The order and structure of *natural laws* suggests a Creator who ordained and conceived those laws. Not erratic, half-baked *chance processes.*

Seriously, from the very beginning of life, supposedly way back when, do you really think we could have survived with half a heart? Half a brain? Half a central nervous system? If the first living creatures were not *fully-formed* **right from the** *beginning of life, they* **could** *never* **survive more than a few** *minutes.* **OK, maybe some people do seemingly function with** *half a brain,* **you got me there. :-) Evolution checkmated once again!!**

A Challenge to Every Evolutionist

Attention: to any evolutionists and/or atheists who want to challenge me on the evolution vs intelligent design (aka: intelligent *purpose*) issue… I will address any issue or challenge you have, ONCE you show me the *transitional* fossils that *must* be found in the worldwide fossil record. You won't find ANY. **Darwinian** (Macro) **Evolution is DEAD!!** It *never* was real, it *never* will be. **This lack of *intermediate* fossils, when there *must* be multiple billions found *already* (*not* in the future), is actually all that is needed to KILL the unsound theory of evolution.**

ATHEISM IS DEAD, DEAD, DEAD!!

Let me make this *Super Simple*… **If *transitional fossils* are *not* found in science museums or in the worldwide fossil record** (and for the 50th time, they indeed are NOT found... ANYWHERE) **- then EVOLUTION IS <u>NOT</u> TRUE. End of fairytale. Meaning that there <u>MUST</u> be an *Intelligent Creator!!* There are NO OTHER (REAL) OPTIONS!! So, if Evolution is *not* true (and, it's *not*)… there INDEED <u>MUST</u> be a GOD, and ATHEISM is DEAD! Thus, ATHEISM is MEANINGLESS and HOPELESS!! But there is HOPE - His name is GOD and He loves YOU!! Surrender your life to Him and you'll see just how amazing He is. There's more on how to do this at the end of this book.**

Why the Genesis Worldwide Flood *Disproves* Evolution

A great flood, the one described in the Bible in Genesis (chapters 6 to 9) - rapidly covered the Earth with water. **It provides a great amount of *evidence* debunking the evolution story, and supports *intelligent creation*.**

First of all, if you're a Christian creationist, but yet believe that a global flood (Noah's flood) did *not* happen, you really need to first read your Bible closer, and then research more if you need to. It happened and there's a *lot* of evidence to prove it.

Let me ask you this, **if you don't believe the Bible's account of the *global* flood, then *what else don't you believe* in the Bible?** *Who* then decides what are made up stories and what is the truth? You? Your mom? Bill Nye the science guy? Anybody else? If *this* is the case, then how could the Bible really be the Word of God?

"ALL Scripture is GOD-BREATHED and is useful for teaching, rebuking, correcting and training in righteousness." - *2 Tim 3*

Keep in mind that the *entire* Bible, not just parts of it, is considered *Scripture, from God to us* (His *love letter to us* so to speak). This verse does *not* say, *"SOME* scripture is God-Breathed..." This Bible verse is either *true* and God-Breathed, or it's *not*. More on this later.

OK, back to the BIG flood... **All across the planet we see *billions* of dead things buried in rock layers, laid down by water. That is *exactly* what a "worldwide flood" *(not evolution)* would do, rather quickly. They could *not* be buried over *millions* of years, because they would be either *scavenged* or *decay* too quickly.**

FAST FORWARD to a modern day similarity… At *Mt. St. Helens in Washington, U.S.A.,* scientists observed the Earth's surface change quite rapidly. Icebergs were buried in hot avalanche material – they melted and **formed badlands in** *DAYS.*

Eruptions in 1980 on May 15th and June 10th produced fine layers in *hours.* Mudflows cut out zigzag canyons 100ft deep in soft sand and mud, complete with perpendicular side canyons. **Canyons that look very similar to those in the Grand Canyon, only on a smaller scale, were produced in…** *HOURS.*

Over the next decade *(not millions of years),* mudflows cut out hundreds of feet of solid rock. *Engineer's Canyon* **was formed in a single DAY by mudflow** – it is 100 ft. deep.

When the global flood happened about 4,400 years ago, sediments of pebbles, gravel, clay, and sand were laid down in successive stratum, covering all animal and plant life on Earth.

Under *extremely great pressure,* **these sediments turned into what we today call** *"sedimentary rock."* Clay became shale; sand turned into sandstone; mixtures of gravel, clay and sand formed conglomerate rock. **All that mass of** *water-laid* **material successively covered multiple billions of living creatures, and plants.**

The result? **Lots and LOTS of fossils, which today are** *only* **found in the sedimentary rock strata.**

When Noah's Flood overwhelmed the world, the first to be covered were *slow-moving* **creatures, the next to be covered were somewhat larger and faster-moving ones, and so it went. Today we can dig into these rock strata and find that the lowest stratum tends to have the slowest-moving creatures; above them are faster ones.**

Evolutionary scientists declare that these lowest strata are supposedly around 570 million years old for the oldest, the *Cambrian,* and the topmost to be the most recent (the Pliocene at 10 million, and the Pleistocene at 2 million years). Creationists say only several thousand.

But, in actuality, we will discover that **the *evidence* indicates that all the sedimentary strata with their hoards of fossils were laid down within a *very* short time.**

Darwin *hoped* (maybe he even prayed) to eventually find evidence of evolution, since he had none. However, it never happened.
Think about this EVOLUTION BUSTER… If evolution were true, we would *not* have to worry about the extinction of species. There would always be *new* ones *evolving.*

The strongest case against the feckless theory of evolution, and worth repeating... just may be that the fossils have been collected for centuries and there are multiple billions in the fossil record. Meaning there *must* be *multiple* billions of *intermediate (transitional)* fossils "found" as well, yet there are none. Make that NONE!! *This fact alone puts an END to the wishy-washy evolutionary theory.* Checkmate for *Intelligent Design* Creation.

NO **ANCESTORS (as in ZERO) have *ever* been discovered** for the massive numbers, possibly *trillions,* of invertebrates (clams, snails, jellyfish, trilobites, etc.), in the oldest fossils found in the Cambrian rock-bed, *anywhere* on Earth.

The fossil record purportedly contains a record of all the billions of years of life on Earth. **If it takes *100 million years* for an invertebrate to evolve through transitional forms into a fish, etc., the fossil strata *has* to contain vast numbers (*multiple trillions*) of the "in-between" forms. But it *never* does!** See the *extreme* *hypocrisy* of the theory of evolution? *It's a sad, uneducated, and not-a-very-funny JOKE!!* - *EvolutionFacts.com (and this book's author)*

"The origin of all diversity among living beings remains a mystery as totally unexplained as if the book of Mr. Darwin had never been written, for *no theory unsupported by fact,* **however plausible it may appear, can be admitted in science."** - *L Agassiz (on the Origin of Species, American Journal of Science 30 (1880), p. 154)*

More Proof for Noah's (Worldwide) Flood

Was Noah's Ark big enough to house all the animals and food? Yes, the ark could *very easily* hold all the different *kinds* of animals on it for a year. As an example, there were only 50 different *kinds* of dinosaurs and the average size was that of a large sheep. They could have easily been *baby* dinos he took onboard.

THE SIZE OF NOAH'S ARK... Based on the Hebrew cubit of 18.5 inches, it has been estimated that **if that great boat, the Ark, was only** *half* **the size stated in Genesis 6:14-16 (omitting water creatures) - it could still have held two or seven of each basic kind of animal and bird. The remainder of the boat was probably used for food storage.**

The Ark was a *barge,* **not a ship with sloping sides, so it had a much larger carrying capacity.** It has been reckoned that, even if measured using the *smaller* 18.5-inch cubit of later times, the Ark would have been so huge that 522 modern *railroad box cars* could fit inside it! One each of every species of air-breathing creatures in the world *today* could be comfortably carried in only 150 box cars. So size was *not* an issue. Size didn't matter.

FLOOD STORIES... Many races and tribes all over the world have, as part of their traditions, stories about a great flood of water that covered the whole Earth. The event was so world-shattering and life-changing that, from parents to children, stories of that great upheaval passed down through the generations. Gradually, as mythologies developed, legends about this flood became part of them. These stories include various aspects of the Genesis flood.

"It has long been known that legends of a great flood, in which almost all men perished, are widely diffused over the world." *- George Frazer (Folklore in the Old Testament, Vol. 1 (1919), p. 105)*

One survey of 120 tribal groups in North, Central, and South America disclosed flood traditions among each of them. *(International Standard Bible Encyclopedia, Vol. 2, p. 822).*

(1) There was general wickedness among men. **(2)** God saw that a flood was necessary. **(3)** One family with eight members was protected. **(4)** A giant boat was constructed. **(5)** The family, along with animals and birds, went into the boat. **(6)** The flood overwhelmed all those living on the Earth. **(7)** The deluge covered all the Earth for a time. **(8)** The boat landed in a high mountainous area. **(9)** Two or three birds were sent out first. **(10)** The people left the boat with all the animals. **(11)** The survivors worshiped God for sparing them. **(12)** A promise of divine favor was given that there would *not* be another worldwide flood of waters.

Another survey of ancient flood literature and legends is discussed by B. Nelson in *The Deluge Story in Stone* (1968). In this tabulation, **the stories and writings of 41 different tribal and national groups** were given. An even larger collection of flood stories is to be found in Sir James G. Frazer's book, *Folklore in the Old Testament* (1919), Vol. 1, pp. 146-330.

NOAH'S NAME... If the story of the Ark and the Flood is to be found among 120 *different tribes* of Earth, should we not expect that Noah's name would be remembered by some of them also Noah's name is found in the stories and languages of mankind. That is another striking cultural *evidence* of the Worldwide Flood which, itself, left so many physical evidences upon our globe. Not only do the rock strata and their fossil contents vindicate the veracity of the flood story, but the languages of man do also! (Much more found at *EvolutionFacts.com)*

"Above all, you must understand that in the last days scoffers will come, scoffing and following their own evil desires. They will say, "Where is this 'coming' he promised? Ever since our ancestors died, everything goes on as it has since the beginning of creation." But they deliberately forget that long ago by God's word the heavens came into being and the Earth was formed out of water and by water. By these waters also the world of that time was deluged and destroyed." - *2 Peter 3:3-6*

Global Flood on Mars?

If there really was a global flood, what would the evidence be? Well, we'd expect to see billions of dead things... buried in rock layers... laid down by water... all over the Earth.

And, you know what? **That's *exactly* what we see!** Billions of dead things buried in rock layers and laid down by water all over the Earth.

DOES THIS MAKE "ANY" SENSE TO YOU?... Many evolution believing scientists believe there was once a *global flood on MARS -* where we *don't* find *any* liquid water (only ice, and in small quantities in vapor in the atmosphere). **Yet they *refuse to believe***

there was a global flood on *Earth* (where they *live*) **which is covered with *SEVENTY* percent *Liquid* Water!** Sure, their beliefs support their evolutionary worldview, but they are *scientific nonsense*.

Doesn't it make you think that if *creationists* starting saying that there "wasn't" a global flood on Earth, that then *evolutionists* would start saying that there *was* one? It seems they want to argue and contradict virtually *everything* in regards to creation. Too bad for them they don't have actual science and evidence on their side, like creation does, in spades. Again, I encourage you to let EVIDENCE (and of course the BIBLE) be your guide to the truth about creation.

The problem *isn't a lack of evidence* for a global flood on Earth. It's a *spiritual* problem. - *AnswersInGenesis.org* (Extra content added by this book's author, as in other citings throughout the book).

The Grand Canyon and Noah's Flood

The Grand Canyon is an outstanding evidence for the Genesis Worldwide Flood.

Evolutionists claim that the Colorado River carved out the Grand Canyon over a period of *millions* of years. Again, another *impossible task* for evolution to pull off, yet evolutionists still claim it as if it actually happened and we just need to believe it. However, let's take a closer look for the truth. **Remember, TRUTH is our quest here, *not* willy-nilly claims and flat out lies with *no actual proof.***

The Grand Canyon is roughly 7,000 feet above sea level at its highest point. The Colorado River enters the canyon at about 2,800 feet above sea level and exits it at about 1,800 feet above sea level... meaning that it would be *IMPOSSIBLE* for it to form the

Grand Canyon, as evolutionists claim, unless it flowed UP. This presents a **huge problem... rivers** *never* **flow up for more than several feet at a time or at any great angle, whatsoever. Water virtually** *always* **flows** *downward* **in the direction of** *gravity's pull.*

The Colorado River lies at the *bottom* of the Grand Canyon, yet it is a typical winding river, the type found in fairly flat terrain. *Winding* **rivers do** *not* **cut deeply!** It is the *straighter, steeper* rivers with swiftly rushing water, that deeply erode soil and hurl loose rocks along their banks downstream.

Notice that the Colorado did little in the way of hurling rocks downstream. This is because the Grand Canyon had not yet hardened into rock when it was cut through. **If the Colorado had carved the Grand Canyon out of solid rock, we would find huge tumbled boulders in and alongside of the stream bed. But this is** *not* **seen.**

Eric Hovind, founder at *CreationToday.org* says... **"The** *EVIDENCE* **does, however, point to** *Noah's Flood.* Today, we see two beach lines from what used to be two large lakes near the Grand Canyon. Creationists believe that after Noah's Flood, the lakes got too full and spilled over the top. When water overflows a dam, the weakest point is instantly eroded. Thus, **the Grand Canyon would have formed** *quickly,* **supporting the** *Biblical Creationist* **flood interpretation perfectly."**

So, **which interpretation is right?** Knowing that *rivers don't flow uphill* and *no leftover sedimentary deposits are found,* evolutionists have a lot of explaining to do when it comes to the Grand Canyon. *The Bible, however, says that a flood covered the whole Earth* (see Genesis 7:18-20). This means *we should find places where the water drained. The Grand Canyon* is one of those places. It is a washed-out spillway that provides solid *evidence* for Noah's Flood."

In 1980, Mount St Helens erupted and created huge canyons with **walls up to *600 feet high.*** One 140 foot deep canyon was later created there in 1982 in **9 HOURS.** Obviously **giant canyons do *not* require millions or billions of years to form,** *only hours or days.*

Don't buy the lame evolutionary thinking theory. Think for yourself friends. The little Colorado River (compared to the gigantic Grand Canyon) did *not* create the Grand Canyon, no matter how much time you give it. It would have been *entirely impossible.* However, a *worldwide FLOOD* could have and in fact, *did* create it, rather quickly, which makes perfect *scientific sense.*

"Living" Fossils

NO CHANGE FROM PAST TO PRESENT... ALL FOSSILS FOUND ARE THE <u>SAME</u> AS THEIR <u>LIVING</u> COUNTERPARTS.

All fossils can be categorized into one of two groups: (1) plants *and* animals which became *extinct,* or **(2)** plants *and* animals which are the same as those *LIVING* TODAY. *Neither* category provides any *evidence* of evolution, because **there are NO TRANSITIONAL FORMS leading up to OR away from any of them.** *ALL* **are only** *distinct species.* **Every last one.**

Some creatures became *extinct* at the time of the *Worldwide Flood* or shortly afterward. But all creatures which did *not* become extinct are essentially *identical* - both in *fossil form* and also in their living counterparts *today!* This is a MAJOR POINT. *No* **species evolution has occurred! None. The fossils provide** *no* **evidence of evolution from one kind to another!** I hope you see that by now.

But then, just as oddly, **the magic *disappears* when the index fossil is found *alive*.** This has occurred with *many* different kinds of species that have been discovered in the past several decades that were formerly considered *extinct for millions of years*, and now found *alive*.

The BIG question to evolutionists is… *Where* then were they all those *millions of years* they were missing from the upper rock strata? Of course they have *no* (real) answer since they are found alive today. Noah's worldwide flood thus makes even more sense.

Apparently, the evolutionary scientists got their original *fossil dating* a bit off *(as in 60 MILLION YEARS off)*, when they found this species of fish ALIVE!! Check this out…

"Most of the species of *maidenhair* are extinct; indeed they served as *index fossils* for their strata until one was found *ALIVE*." The youngest fossil, *Coelacanth*, is about *sixty MILLION years old*. Since one was rediscovered off Madagascar *(found in 1938)*, they are no longer claimed as 'index fossils' - which are fossils which tell you that all other fossils in that layer are the same ripe old age." - *Michael Pitman (Adam and Evolution (1984), pp. 186, 198)*

Many more *(so-called extinct species)* have been found *ALIVE* since then. The funny thing is that **they look the *same* NOW as their alleged 6 to 50+ *million* year old great, great, great, etc. grandparents.** Huh? **NO evolutionary changes in 50 MILLION years?? Something seems pretty fishy. Yet, evolutionary theory "stories" that *they became extinct way back when*, have caught many people hook, line and sinker.**

In reality, within *each stratum* is to be found an utter confusion of *thousands* of *different* types of plants and animals. The evolutionists maintain that if just one of a certain type of creature (*index fossil*) is found anywhere in that stratum, it must automatically be given a certain name, and more: a certain *date* millions of years ago when all the creatures in that stratum are supposed to have lived. Just more craziness we're all supposed to buy into. Don't do it.

Yet, just by examining a particular index fossil, there is *no way* to tell that it lived many millions of years ago! It is all part of a *marvelous imaginary theory (fairytale),* which is actually nothing more than the *GRAND EVOLUTIONARY HOAX.* Experienced scientists *denounce* it all as completely *untrue.* A full blown *SHAM* indeed. I like to call it a fairytale for itching ears.

Any rock containing fossils of one type of trilobite *(Paradoxides)* is called a *Cambrian* rock, thus *allegedly* dating all the creatures in that rock to a time period *allegedly* around 600 million years in the past. But rocks containing another type of trilobite *(Bathyurus)* are arbitrarily classified as *Ordovician,* which is claimed to have *supposedly* spanned 45 million years and begun 480 million years ago.

How can anyone come up with such ancient dates simply by examining two different varieties of trilobite? The truth is that it *can't* be done. It's *Science Fiction* to even pretend to do so. *The evolutionism fairytale plot thickens even more.*

Add to this... the problem of mixed-up *index fossils* – This is when index fossils from *different levels* in the strata are found *together!* And, there are plenty. This is a *HUGE problem* which for obvious reasons, paleontologists do *not* publicly discuss. They can't of course, for fear of destroying their precious, incompetent little theory they so dearly cling to like *"ugly on an ape."* (an old saying for you new young friends).

Speaking of apes (This is this book's author, Nick, speaking. Please excuse my *rabbit trail* for a minute)... Did you ever wonder why *parrots* **can talk** (mimic us), and yet *primates can't?* Maybe it's a fun little joke God plays on the evolutionists... ***"See there my non-believing friends on Earth, the parrot talks, but I kept your prize possessions mute. Why do you think that is? Shouldn't the monkeys be the talking animals? I'm just trying to get your attention, friend. Stop all the monkey business and let's talk soon."*** *- Love, God*

As we analyze one aspect after another of all kinds of evilution, oops, I mean evolution *(stellar, geologic, biologic, genetic, etc.),* we find it all to be little more than a carefully contrived *science fiction story.*

The truth is that the evolutionary *Tree of Life* is just another *fake,* like all the other *so-called evidence* for the blemished evolutionary theory.

Without *Transitional Fossils*, there is <u>*NO EVOLUTION*</u>, because **THAT is what evolution is ALL about my friends.** No Transitional Fossils = EVOLUTION LOSES!!

Evolutionists DATE FOSSILS BY A THEORY... But now comes the catch: ***How can evolutionary geologists know what DATES to apply to those index fossils?*** The answer to this question is a wild, uneducated *GUESS,* something they like to call a *theory!*

Here is how they do it: Darwinists theorize *which* animals came first, and *when* in history they appeared on the scene. Then they date the rocks according to their powerless theory - *not* according to the wide mixture of fossil creatures in them - but by assigning dates - based on their *theory,* according to certain *index* fossils. **That is a Huge, *Circular-Reasoning* Hoax!** *- EvolutionFacts.com*

Circular Reasoning

When we examine it, we find that the **strata-dating theory** is based on **no more** than **circular reasoning.**

Circular Reasoning is a method of *false* logic... by which *"this* is used to prove *that,* and *that* is used to prove *this."* It's a joke within science actually, but sadly some people still choose to believe whatever they're told. **That's not you, right? You question this stuff and demand proper answers that provide actual *evidence,* right?**

There are *several types of* **Circular Reasoning** found in support of the insufficient evolutionary theory. One of these is the hilarious *geological dating position* that says - **"fossils are dated by the type of stratum they are in, while at the same time the stratum is dated by the fossils found in it."** **SAY WHAT??** Are you drunk again? Please repeat that, I must have heard you wrong.

The funny thing is that science textbooks actually say these harebrained things and people don't think twice. They just gobble it up as if it's the gospel truth. *"Feed me more of that evolutionary candy."* Yeah, maybe it's the *evolutionary gospel according to Uncle Charlie and the gang,* and call it science if you will, but it's not *real* science by any stretch of the imagination. **It's *imaginary science.***

An alternative evolutionary statement is that, *"the fossils and rocks are interpreted by the theory of evolution, and the theory is proven by the interpretation given to the fossils and rocks."* You're thinking *double talk* too, aren't you? Well, *correct-amundo.*

Evolutionists - **(1)** use their theory of rock strata to date the fossils, **(2)** and then use their theory of fossils to date the rock strata! An example of *Circular Reasoning* at its finest.

The professor of paleobiology at Kansas State University wrote this about the *problem* of *circularity:*

"Contrary to what most scientists write, the fossil record does *not* support the Darwinian theory of evolution, because it is this theory (there are several) which we use to interpret the fossil record. By doing so, we are guilty of *Circular Reasoning* if we then say the fossil record supports this theory." *- Ronald R. West (Evolutionist. "Paleontology and Uniformitarianism," in Compass, May '68, p. 216)*

A Double Circle

Circular Reasoning is the basis, not only of the fossil theory - but of the *entire* theory of evolution!

First, reasoning in a circle is the basis of the (so-called) *evidence* that evolution has occurred in the past. *(The fossils* are *dated by the theory of strata dating; the individual stratum are then dated by the fossils).*

Second, reasoning in a circle is the basis of the *mechanism* by which evolution is supposed to have occurred. (The survivors survive. The fittest survive because they are fittest - yet, according to that, ALL they do is <u>SURVIVE</u>! They do <u>NOT</u> EVOLVE into something better!)

By the use of *circular reasoning,* evolutionary theory attempts to *separate itself from the laws of nature!* Limiting factors of chemical, biological, and physical law *forbid* matter or living creatures from originating or evolving.

Actually, the entire theory of evolution is based on one vast circularity in reasoning! Because they accept the theory, evolutionists accept all the foolish ideas which attempt to prove it.
- EvolutionFacts.com

DINOSAURS and HUMANS
(Proof They Lived Together)

Evolutionists claim that dinosaurs are between 65 million and 135 million years old. Yet, UniversityToday.com says the modern form of humans only evolved about 200,000 years ago.

In contrast, the vast majority of *creationists* agree with the Bible and say mankind is only about 6,000 years old (which by-the-way, for reference, is older than the oldest man made structures, such as the *great pyramids*. Curiously, **we have *never* discovered any man-made structures older than about 4,500 years or so,** yet we're told by evolutionists, that humans date back to 200,000+ years old).

Truth told, the *facts prove* **that Dinosaurs are NOT millions of years old**... In fact, there's **tons** of real evidence that **humans and dinosaurs** *lived together* **at the** *same* **time in history.**

Every major group of vertebrates has been found in or close to dinosaur fossils. If evolution is true, then dinosaur fossils should primarily be found by *themselves.* - *Institute of Creation Research (Guide to Dinosaurs, 2013, pg. 56)*

Do you think that **Marco Polo** would have journaled about huge dinosaurs (in detail) back in the 13th century if he never saw them? I highly doubt it. WHY would he? The quote below sounds like a T-REX possibly, and it's a far cry from being 65 million years old.

They were called *dragons* **back then. For reference, the word** *dinosaur* **didn't come into being until 1841** by biologist and paleontologist, Sir Richard Owen. The word dinosaur derives from two greek words and means "terrible lizard."

***Marco Polo* wrote about seeing *large dragons* (dinosaurs) during his travels in the late 1200s.** He journaled: *"Here are seen huge serpents, ten paces in length (about 30 feet), and ten spans (about 8 feet) girth of the body. They have two short front legs, having three claws like those of a tiger, with large glaring eyes. The jaws are wide enough to swallow a man, the teeth are large and sharp, and their whole appearance is so formidable, that neither man, nor any kind of animal can approach them without terror." - Forbidden-History.com*

Fossilized footprints of dinosaurs *and* humans have been found *together* in places like *Glen Rose, TX* (where a human footprint was actually *inside* the dinosaur footprint fossil), as well as *several other locations* around the globe. **This proves that they lived at the <u>same</u> time,** *not* 65 to 135 million years apart. *- AnswersinGenesis.org*

BUSTED... "Around 1977, *evolutionist* Mary Leaky (wife of famous anthropologist, Louis Leakey) **found in Africa, *human footprints*** in strata that *evolutionists* date at nearly **4 million years old. They are <u>*identical to modern human footprints*</u>. These and other footprints disprove evolutionary theories, especially those in which dinosaur prints are found with human footprints. Dinosaurs are said to be dated from 65 million to 135 million years ago; whereas man is said to have appeared far more recently."** *- National Geographic, April 1979; Science News, February 9, 1980.*

Evolutionists parade dinosaur bones as a grand proof of evolution, when they are *not* proof at all! ***Extinction* is *NOT* evolution as some naively suggest.** *- EvolutionFacts.com*

Over the years, *Creation Magazine* has highlighted *sculptures, paintings, tapestries, vases, ornaments, etc.,* where people have depicted *dinosaurs* that they must have seen *firsthand,* because of the incredible detail in the representations. These go back before

the *13th century*. **There are now many examples like this found all around the world, an evolutionary *impossibility* if man and dinos are separated by 65-135 million years.** - *Creation.com*

DINOSAUR "SOFT TISSUE"... Another compelling reason why dinosaurs *can't* possibly be millions of years old is *soft tissue* **found in dinosaur fossils.**

Creation Magazine broke the news back in the 1990s when **paleontologist *Dr Mary Schweitzer* (evolutionist) claimed she found *unfossilized blood cells* and *blood vessels* in the leg bone of a supposedly 65 million year old *T-REX*.** At the time there was a massive effort to discredit the claim, trying to prove they were anything but dino blood cells (*as this would further help destroy the evolution theory*). How wrong the critics were! **We now have *numerous* examples of *soft tissue* finds, including - ligament, flesh, more blood cells, and even *skin*.** These are all featured in *Creation* Magazine (*my favorite magazine*) and at CREATION.com.

Dr Mary Schweitzer's discovery *doesn't* fit the evolutionary story. It *does* support the biblical account. Starting with Genesis, we know God created everything - including dinosaurs - in six days, just thousands of years ago. - *Ken Ham (AnswersinGenesis.org)*

Of course, if fossils are traces of life that lived a million or more years ago, then we have *no* reason to believe original organic molecules should be preserved, let alone cells or whole tissues like blood vessels. But *many* fossils (not only the T-REX) do have these stunning features. - *AnswersinGenesis.org*

During his research at the Paluxy River Bed near Glen Rose, TX, **Dr. Bird found not only *human* footprints, but also, by them, trails of *large three-toed carnivorous dinosaurs*, and the tracks of a *gigantic*** *sauropod.* Each print was a huge 24 x 38 inches [60.9 x 96.5 cm] in size, 12 feet [36.57 dm] apart, and sunk deeply into the mud! Both man and dinosaur were apparently running. Folks, these are *real* dinosaur discoveries housed in the great state of Texas for your examination. *- EvolutionFacts.com*

"Evolution is *unproved* and *unprovable,* we believe it because the *only* alternative is special creation, which is unthinkable." *- Sir Arthur Keith (a militant Anti-Christian physical anthropologist)*

Cryptozoology... No, this is *not* some new Bitcoin type of *crypto-currency.* Cryptozoology is the study of *hidden animals. Cryptozoologists* are the people who believe there may still be huge dinosaurs and creatures living among us today (We are not talking about known big lizards and alligators). This is an attempt to prove that creatures widely regarded as extinct or imaginary are actually alive but merely hidden. Such creatures are called *cryptids.* Yes, it's possible, however, a lot of searching has been done over the past several decades or so, and even though there are rumors of them in remote places, it appears that nothing really has been discovered yet.

Thousands of Years OR 180 Million?

Shells **from as far back as the *Jurassic stratum,*** which **supposedly dates back to *135-180 million years ago,*** have been **found to have** *amino acids* **still locked into protein structures.** The amino acid residues came from *inside* those shells - **so the shells *cannot* be more than a** *few "thousand" years old.* This is just one example of many others found around the world.

So, who's correct, the evolutionary scientists who are making a *wild* **GUESS**, or the *scientific* **FACTS**?

The Most Well Known
Evolutionary HOAXES

Let's look at *supposed* intermediate *(transitional) fossils* that have been proven either *hoaxes,* or something *other* than actual transitional fossils. (Keep in mind, even *if* they had *not* been proven hoaxes *to fool the public,* **there** *must* **be multiple** *billions* **of transitional fossils** *discovered* **for evolution to be true).**

ALSO… *if* the transitional changes occur over millions of years, then *there should be vast numbers of transitional species LIVING today,* as well as etched into the fossil record. But they are nowhere to be found. They *don't* exist and *never* have existed.

All of the biggest alleged evidences for evolution have been exposed as hoaxes, yet a few *still remain in museums today.* **What?** I call SCAM.

This shows how desperate or greedy some people can be. Some desire fame, some fortune, some both. Some just don't want to be accountable to their creator. None were contrived out of ignorance, but rather, (sort of) well thought out schemes to fool gullible people who believe whatever they are told. Please tell me this *isn't* you.

HERE'S THE SKINNY… Even if any of these *frauds* mentioned were true (they are *not,* but for argument's sake let's say they were), they would be a colossal joke. A small hand full of fossils said to be *transitional* (intermediate) fossils, from one species to another different kind of species (dog to bear, fish to reptile, reptile to bird, molecule to man, etc), would *not* prove anything. *In reality…* with 100 million to 200 million fossils in the museums of the world, and 100's of billions to trillions in the fossil record… **Billions of** *transitional* **fossils** **MUST** **be found in these museums and**

hundreds of billions to even trillions in the fossil record – that is, if evolution were true. **However, there are *none* in existence. Let me repeat that, there are… <u>NONE</u>!!**

This enormous fact alone *completely disproves* **the frail theory of evolution…** and no excuse or argument from anyone can combat against it. **The transitionals** *had* **to be there and** *must* **have been found, but were** *not.* We still should be finding them today, but we are *not.* To say otherwise is complete ignorance or deceit. Do not be duped by any other answer. That would be *Fake News* as well.

The following is one of the *strongest, indisputable evidence AGAINST evolution!* **It's** *not* **possible to disagree with it. In fact, it** *completely* **annihilates the theory!!** Bye Bye Evolution!

There Should Be <u>NO</u> Species
(If Evolution Were True)

IF EVOLUTION WERE TRUE… there could *NOT possibly* **be** *ANY* **distinct species at all! There would** *ONLY* **be** *INNUMERABLE TRANSITIONS!!* Meaning there would *only* be a *confused blur of transitional forms*, each one only *slightly different* from the others.

This is a very significant and important point. Categories of plants and animals can be arranged in orderly systems *only* because of the separateness of the species. - *EvolutionFacts.com*

"Professing to be wise, they became fools, and changed the glory of the incorruptible God into an image made like corruptible man, and birds and four-footed animals and creeping things. Therefore God also gave them up to uncleanness, in the lusts of their hearts,

to dishonor their bodies among themselves, who exchanged the truth of God for a lie, and worshiped and served the creature rather than the Creator, who is blessed forever. Amen." - *Romans 1:22-25* (The Bible)

The Popular Hoaxes

Following are the majority of the most popular hoaxes that money grabbing, fame seeking shysters have tried to pull off in the past 100 years or so. Every one of them has been busted as a fraud. Keep in mind, for evolution to be true, there would have to be millions, even billions of *transitional* fossils in the fossil record, yet we find none.

Archaeopteryx

The word means, *Early Wing.* Some evolutionists claimed Archaeopteryx was a transition between dinosaurs and birds. It has been proven to simply be a *bird.* However, some scientists even believe it was a *hoax,* where the wings were added to a land animal's body. Either way, **it has been proven *not* to be a transition between dinosaurs and birds.**

I Love Lucy

Lucy (Australopithecus)... **One of the most famous of all so-called** *evidence* **for evolution was a** *chimpanzee* **named** *Lucy.* She was *made* to look more human than ape, thus trying to *fool* people into thinking she was a transition between apes and humans.

She has been *proven* **to be - 100% APE.** Her skull was like that of an ape, her brain was the size of a chimpanzee brain, her thumbs were exactly like that of an ape, here toes were long and curved for climbing trees, her knee and elbow joints were those of a primate, she had massive V-shaped jaws *unlike* man, she had short legs just like a chimp, and her ankle bones tilted *backward* like an ape (ours tilt *forward*).

Even the man who discovered Lucy's fossils in 1975 in Ethiopia *(while listening to the Beatles song, Lucy in the Sky with Diamonds),* **later** *admitted she was only an ape.* According to experts, she was probably just a chimpanzee.

The funny thing is that even though she's been *proven a fake,* we still see stuffed replicas of her in museums around the world to this day. Evolutionists just being true to the scam I guess.

Nebraska Man

Nebraska Man **was used in the famous** *Scopes (Monkey) Trial* **in** 1925, in which the evolutionists had a huge (short lived) public victory. Shortly after, it was learned that **an artist had been hired to draw a** *picture* **of a** *prehistoric man and his entire family,* **from, get this, one** *tooth* **that had been discovered. ONE TOOTH. Turns out it was just a** *pig's* **tooth.** Another evolution hoax busted as *fraud.*

Piltdown Man

Evolutionist's Piltdown man is now universally known to be a **deliberate hoax, consisting of only an** *ape's* **jaw and** *human* **skull.** According to History.com, *carbon dating* says it's not 50,000 to

500,000 years old as claimed, but actually less than 600 years old. **The ape's bones were stained and it's teeth were even filed down and stained to appear to be human.** - *EvolutionFacts.com*

"If I knew of *any* **Evolutionary** *transitional's*, **fossil or living, I would certainly have included them in my book, 'Evolution.'"** - *Dr. Colin Patterson (Evolutionist and senior Paleontologist at the British Museum of Natural History, which houses 60 million fossils)*

Neanderthal Man *(100% Human)*

To say *Neanderthals* were merely *subhuman* or a *transition* between ape and human is ridiculous and ignorantly perverts the actual *facts.* A lot of *evidence* proves they were *fully human.*

Even Thomas Huxley, the vehement evolutionist and defender of Charles Darwin, said that the Neanderthal bones found in Neander Valley in Germany, belonged to *humans*.

Many evolution and creation scientists agree that the deformed bones were from *modern* **humans** *(homo sapiens)* **stricken with** *arthritis* **and** *rickets,* **due to lack of sunlight.** This was the result of worldwide volcanic dust pollution during the *ice age,* forcing them to spend much of their time in dark, damp caves.

FAKE NEWS Report... **Historical** *evidence* **shows that shady evolutionists had intentionally** *mismatched* **the upper and lower jaw, in order to make the Neanderthals look more like apes.**

Check this out... **A Neanderthal skeleton was found in Poland in 1908, buried in a suit of chain armor that was not even completely rusted.** Think about this - *NOT all rusted,* yet evolutionists say they lived 200K to 300K years ago. - *"Neanderthal in Armor" (in Nature Magazine, April 23, 1908, p. 587)*

Think about it, we still see living apes today. We see man. **Where then are the *transitional beings* that should be *living* among us today?** Not even one? Heck, they can't even produce one *intermediate fossil.* Yeah, solid theory. Right.

IN A NUTSHELL... Neanderthals *made cave paintings, jewelry, fires, used cooking utensils they made for the food they gathered or hunted with weapons they crafted, buried their dead, walked upright like modern man, and had bigger brains than ours.* Maybe modern day humans are the real *pea-brains.*

Neanderthals were 100% Homo-Sapien HUMAN. End of Story.

How About *Cro-Magnon* Man?

Just 12 years after the first Neanderthals were discovered, a number of skeletons were found in a cave in France in 1868. *Cro-Magnon* man was discovered and also considered the link between ape and man. (Cro-Magnon means *Big Hole*).

The problem is... **their brains were bigger than modern day humans. They were excellent artists drawing and painting many colorful cave murals, made musical instruments, sewed clothes, cooked, baked pottery, and even kept astronomy records. Some were over 6 feet tall.**

And, of course **"no" transitional fossils have ever been found that would actually link ape to man.** Just another fairytale told by creative storytellers of the day.

Java Man

In 1891, *Java Man* was discovered on the island of Java (Dutch East Indies, now part of Indonesia). Evolutionists estimated it to be between 700,000 and 1,000,000 years old. **Several years before his death, the man who discovered *Java Man* admitted it was only the bones of a *Gibbon* (a kind of ape).** Funny thing is that a *Starbucks coffee mug* was found next to the body, thus proving even primates like *coffee. Ha,* I just wanted to see if you're paying attention. :-)

Are Man and Ape First Cousins?

Charles Darwin did *not* include his viewpoint on *ape to human evolution* in his book, the *Origin of Species,* in 1859, but *did* in another book 12 years later.

Darwin lamented, "Why, if species have descended from other species by fine gradations, do we *not* everywhere see *innumerable transitional forms?"* Darwin assumed that this lack of fossil evidence was a result of the incompleteness of the fossil record in his day and predicted that as more fossils were discovered, the *"missing links"* supporting his theory would eventually be found.

There is no scientific reason to doubt the literal truth of Genesis 1–5 about the origin of humans (or the origin of anything else in those chapters). - *AnswersInGenesis.org*

Funny thing is that many evolutionists tell you they know everything about the *missing links,* except for the fact that **they are all *MISSING.***

There are *many* differences between humans and apes such as: adult apes have brains that are generally about one-third the size of normal adult human brains, the shape of the skull is *drastically different,* their hands, feet and many other body parts are *very different,* recent scientific studies have shown our DNA is actually *vastly different* than what evolutionists had been saying. Humans talk, read, sing and create, which obviously apes do *not* do very well. In fact, have you ever heard a monkey singing, reading poetry, playing Minecraft, making breakfast, driving a car, or creating anything other than a big hot mess?

Maybe the biggest proof *against* **"ape to human evolution" is the complete** *lack of evidence* **in the** *fossil record.*

ISN'T IT FUNNY...   Isn't it funny that from birth we're all encouraged to *question* everything under the sun, EXCEPT, the things of evolution, like... "How did everything come from nothing?" Or, "Where are the transitional fossils that prove evolution?" These are off-limits. We're just supposed to blindly believe them. I refuse. How about you?

PLEASE DO ME A FAVOR and show some love… post some of these topics on your Facebook, Instagram, and Twitter pages and link them to: facebook.com/CharlieWasWRONG). Do it often. Email and text to your friends, too. You can always find more on our facebook page (@CharlieWasWrong), and in the *ebook* from Amazon at CharlieWasWrong.com.

CHARLES DARWIN
(The Man, the Myth, and the Legend... in His Own Mind)

Mr. Darwin did not know of even ONE instance where any species changed into a different species (cat to dog, fish to reptile, etc).

"Not *one* change of species into another is on record . . . we *cannot* prove that a single species has been changed." *- Charles Darwin (Life and Letters)*

The same still holds true *today,* not even *ONE example of any species evolving into another has been discovered!!* **Surely not for lack of trying or falsifying. Evolution is *false presuppositions* at their finest.** If you think this is incorrect, *PROVE ME WRONG.* I double-dog dare you. A big dare indeed. But before wasting too much time, keep in mind that no evolutionary scientist has *ever* been able to find any (real) *transitional* fossils. Yet, as mentioned earlier a few times and worth repeating - there *must* have been *BILLIONS* of *transitional fossils already* discovered for evolution to be true.

Charlie relied heavily on MADE-UP *STORIES* to build upon his fictitious theory of evolution, <u>NOT</u> *FACTS!!*

Let's take a little closer look at some verifiable *facts* (not stories) on the Charles Darwin most people never hear about. **Was Charlie the great discoverer of biological evolution, or was he actually a *Plagiarist, Racist,* and *Convincing Liar?***

Ruffled some feathers there I'll bet. Don't take it out on the messenger, rather blame the theory that has lied to you all your life. This madness can stop anytime now. If you disagree, then *prove* that he wasn't a *plagiarist, racist, and grand storyteller.*

Most people know that Darwin is best known for teaching that man evolved from common ancestors, primarily monkey to man. This wasted theory has since *evolved* with many modern evolutionists (with absolutely *no evidence* of course) who now claim that man *and* ape both descended from a *Common Ancestor.* However, both are provable *false* statements.

Darwin taught that humans are *not accountable for their actions* since we *evolved* by *pure chance.* So in essence, we are all basically, somewhat tamed, wild beasts. Life is a battle of the *"survival of the fittest,"* so to speak. It's a *dog-eat-dog* world.

Keep in mind, **Mr. Darwin (*not* doctor or sir) was *not* a college educated scientist with a doctorate**, but rather an educated man that just so happened to teach a false theory that he believed true at the time, although *he had serious doubts about his speculative theory of evolution,* as he mentions in a few of his later books.

We all know that ***believing* something is true doesn't necessarily make it true, no matter how much we may consciously or subconsciously *want* it to be. Real *evidence* needs to be the deciding factor for naturalism, *not a man's opinion* or the ability to write good *science fiction.***

Empirical Science (science that can be *observed*) has blown the evolutionary hypothesis away over the past century or so. Many evolutionary scientists absolutely know that this deficient theory is *chock-full of holes,* yet they still subscribe to it for different personal reasons. Perhaps they simply *refuse to believe there is a supernatural*

God who created the universe and all life, a God *outside* of time and space, so to speak. They would rather deceive themselves and believe that they are in charge of their lives.

Uncle Charlie frequently commented in private letters *(that you can find online and in books and articles)* that he recognized that there was <u>*no evidence*</u> *for his theory (invented stories),* and that it could destroy the morality of the human race. It certainly hasn't helped.

**BESIDES LINKS, WHAT ELSE
WAS DARWIN MISSING?**
HIS MARBLES.

Darwin *(The Plagiarizer)*

Fighting words to evolutionists, but truth is truth. Evolutionists are not going to like their golden child being exposed like this but people need to know the *truth*. You do want to know the truth, don't you?

Some historians believe that all of the major contributions with which Darwin is credited in regard to evolutionary theory, including *natural selection,* actually were *plagiarized* from *several* other biologists and thinkers up to 100 or so years *earlier.*

Many, if not most, of Darwin's major ideas are found in several *earlier* works. Charlie rarely gave due credit to the many people from whom he liberally *"borrowed"* material for his book.

Furthermore, most (if not all) of the major ideas (*including Natural Selection*) credited to Darwin actually were discussed in print by *others* a *long* time before him. Thus, there is a lot of *proof* of his plagiarizing that can be researched.

Actually, proposals that one type of *animal,* **even** *humans,* **could descend from other types of animals, are known to go back to the first pre-Socratic Greek philosophers, about 500 BC. That's over 2,000 years** *before* **Charlie was born.**

One of the most important *pre-Darwinists* **was Charles Darwin's own** *grandfather, Erasmus Darwin.* He discussed his ideas at length in a popular two-volume work, ***Zoonomia,*** published in 1794. Apparently, Erasmus Darwin *originated almost every important idea that has since appeared in evolutionary theory,* including *natural selection.* UNLESS he *too* plagiarized it.

Even though large sections in many of **Charlie's books** *closely* **parallel his grandpa's writings,** Darwin *never* once openly admitted that his grandfather had a major influence on his central ideas. - *Jerry Bergman (creation.com/darwin plagiarism)*

Alfred Russell Wallace... a known Marxist and spiritist (like many evolutionists of the day) is **considered to be the man who** *developed the Theory of Evolution* **in his** *Ternate Paper,* which *Darwin pirated and published* under his own name. Darwin, a wealthy man, thus obtained the royalties belonging to Wallace, a poverty-ridden theorist.

Wallace conceived the idea of the *"Survival of the Fittest,"* *(Darwin simply stole it)* as being the *method by which species change.* Ironically, the concept proves <u>nothing</u>. *"The fittest? Gee, which one is that?* Duh - *It's simply the one that survived longest. Which one survives longest? The fittest."* This is the epitome of *Circular Reasoning.* **The phrase says** *nothing* **about the evolutionary process, much less proving it.** - *EvolutionFacts.com*

Mr Wallace eventually *separated* **from Darwin's position** - the position he had given Darwin - when he (Wallace) realized that **the**

human brain was *far too advanced* for evolutionary processes to have produced it. *(Loren C. Eiseley - "Was Darwin Wrong about the Human Brain?" Harper's Magazine, 211:66-70, 1955)*

Did Charlie's Theory of Evolution *Plagiarize God?* - Sure, Darwin spun many of the past findings on the shakey theory of evolution in his own words, and blatantly plagiarized some… which helped to make him the popular *professed father of evolution.* **He simply *plagiarized* and *popularized* it,** so that's really all he should get any real credit for. But to go deeper, you could even say that **the *fable of evolution* has basically *plagiarized God's Creation,* and changed it to fit its own twisted fantasy.**

* Details of this information (and most everything else in this book) can easily be found online, in books, or in magazines if you want more insight. (I, the author, often add comments. I just can't hold back. I don't always remember to mention that I added the snarky comments. Oops)

Darwin *(the Racist)*

Most people don't know **the *full* title of Charles Darwin's most well known book,** which is…

"On the Origin of Species by Means of Natural Selection, or the Preservation of *Favored Races* in the Struggle for Life."

You read that right… "FAVORED RACES." Favored meaning "preferred." If you don't believe me, look it up. That's the entire title of his book. The last half of the title, usually overlooked, sounds like it could come straight out of a *Ku Klux Klan* manual.

Virtually all of the world's *mass murderers* became strong proponents of evolution. They felt humans were no more special than animals, especially certain human races, unborn babies, and people with disabilities. **In their sick and demented minds, the theory of evolution helped them justify mass murder.**

Get this... In his famous book, *Origin of Species,* **Darwin *never* did explain the *ORIGIN* of Species.** What?? How ironic. **The title is about how species *began,* and then he *never* tells us how they first actually *originated.*** Why don't more people question this?

BUSTED by the NUMBERS... Back *in 1859 there were less than 15% of the now estimated 1.6 million species* (however, many scientists claim there are at **least *10 million* living species today,** which would only make Darwin's theory much less credible at just 2.4%). **Darwin was *not* even aware of 85% to 97.6% of the different species. How would having only 2.4% to 15% of the so-called evidence, go over in a *court of law* today?** Yes, a valid question if we are to keep a proper perspective.

Keep in mind, it *wasn't* even evidence, but rather just a *theory.* That makes the case even weaker. See how far you get with Judge Judy (or any judge and jury) with less than 15% of the *hypothetical* evidence on *any* topic, let alone one with *no real evidence, just hearsay.* Also, keep in mind that **DNA was not discovered until almost 100 years *after* Darwin's theory.** When studied, DNA proves conclusively that the unsound theory of **evolution has always been a complete *sham.*** Anybody believing differently is accepting a bunch of half-truths (and all out lies), instead of examining all of the real scientific *facts!*

"I am *not* satisfied that Darwin proved his point or that his influence in scientific and public thinking has been beneficial - the success of Darwinism was accomplished by a *decline in scientific integrity.*" - *W.R. Thompson (Evolutionist. An Introduction to Charles Darwin, Origin of the Species)*

Darwinians (aka: Darwinists)
Who Were... *Mass Murderers*

"**The first point is that selfishness and violence are inherent in us, inherited from our remotest animal ancestors. Violence is, then, natural to man, a product of evolution.**" - *P.J Darlington (Evolutionist. "Evolution for Naturalists" - 1980 - pp. 243-244)*

The idea that man is but a beast, and *not* accountable for any of his actions, is the heart of Darwin's teaching, and unleashes the *worst* in man.

For the record... much historical *evidence* points to the fact that ***tens of millions* of children and adults, *and* two *Billion* unborn babies, were inhumanly murdered in the last 50 years. TWO BILLION!! Largely due to the *Darwinian (including its racism and eugenics)* beliefs strongly held by *Marx, Lenin, Stalin, Hitler, Mussolini, Mao, Nietzsche, Margaret Sanger, Abortion Doctors,* among others.**

Currently, over *55 million babies are brutally murdered* in their mother's bellies *every year* around the world. **What has this world come to when a mother's *rights* are considered more important than a defenseless human baby's *life?* We live in such a SICK and DEMENTED world!!** Try killing a baby eagle once. You'll go to jail and not pass go and collect $200.

If you doubt this information, you can easily find an abundance of *factual proof* illustrating their horrific Darwinian beliefs - all over the Internet, in magazines, newspapers, and books.

As an example, **Nazism is based on the philosophy of evolution.** You can read all about Hitler's racist philosophy in his book, *Mein*

Kampf. He thought that the *Germans were the superior race* that deserved to rule the world and that everyone else was *wasting space.*

A line runs from Darwin to Francis Galton (the father of *eugenics* and *Darwin's cousin*), **to 11 million innocent people (6 million Jews) being brutally tortured and murdered in Nazi Germany under Hitler's reign.**

Specialists in Hitlerian studies note that *Hitler hated Christianity as fiercely as he loved Darwin's theory.* He believed he was carrying Darwinism forward with his doctrine that undesirable individuals and inferior races must be eliminated.

"Atheism, *not* religion, is the real force behind the mass murders of history." - *Dinesh D'Souza*

Yes, Hitler hated Christians and was indeed a big believer in Darwin's flawed theory of evolution, and so were virtually all *other* mass murders in history. Their reasoning was, "if we all just evolved by *purposeless chance* from primates, what value do we have anyway?"

Actually, God thinks about *you* much differently. In fact, he loves and values YOU very much. You mean the world to Him.

"The German Fuhrer (Hitler), as I have consistently maintained, is an evolutionist; he has consistently sought to make the practices of Germany conform to the theory of evolution." - *Sir Arthur Keith (Militant Anti-Christian physical anthropologist)*

Darwin's Little Finches

So, what's the big deal about Darwin's finches, anyway? Actually, *nothing* when you know the facts. For over a century, many school textbooks have used the Galapagos Island finches (also called, *Darwin's Finches*) as a primary example of *evolution-in-action.*

THE FACTS ARE - that the differences between these finches are *far less* **than the differences in the human population – and virtually** *nobody* **claims that humans are still evolving today!**

Keep in mind, the important thing is that the *finches always remain finches.* They never have evolved into other species, never will, and never can. Never!! Ever!! **Just like** *every* **other animal and plant has always remained true to its kind.** An enormous and obvious evolutionism killer all on its own.

Within 14 years after writing *Origin of the Species,* Darwin confessed to a friend:

"In fact, the belief in Natural Selection must at present be grounded entirely on general considerations [faith and theorizing]. When we descend to details, we can prove that *no* **one species has changed -** *nor* **can we prove that the supposed changes are beneficial, which is the groundwork for the theory. Nor can we explain why some species have changed and others have** *not."*
- *Charles Darwin (letter to Jeremy Bentham, in Francis Darwin (ed.), Charles Darwin, Life & Letters, Vol. 3, p. 25)*

Evolutionists who subscribe to the evolution theory insist that the *environment* caused these differences to develop so quickly. Hey, **what happened to the** *"millions of years"* **part of the theory?** Huh, guess they forgot about that for a minute. Brain fart.

Creationists would agree but do *not* call it evolution. Rather, we call it... **"ADAPTATION." Simply giving one of these** *variations another species "name,"* **does** *not* **make it a different species. In fact, to go as far as claiming it's proof for** *evolution-in-action* would be *scientifically irresponsible and incorrect.* **In other words,** *completely ridiculous, and then some!!*

Scientists acknowledge that all dogs descended from a common ancestor, and all are still *DOGS.* Yet **there are** *far greater differences* **among dogs than there are among Darwin's finches** or most other *sub-species* in the world. All biologists classify dogs as being in the *same* species. - *EvolutionFacts.com (NOTE: A couple lines or so have been added by this book's author, to some of the EvolutionFacts.com information cited in this book).*

"Unfortunately for Darwin's future reputation, his life was spent on the problem of evolution which is deductive by nature . . It is absurd to expect that many facts will *not* **always be irreconcilable with any theory of evolution; and, today,** <u>*every* **one of his theories is** *contradicted by facts*</u>**."** - *T. Mora (Evolutionist. The Dogma of Evolution, p 194)*

Q) DARWIN'S UNDERSTANDING OF BIOLOGY QUALIFIED HIM
FOR WORK IN WHAT TYPE OF CELLS? A) PADDED CELLS.

Evolution Scientists Say Darwin's Theory Is *Unscientific* and *Worthless*

Here's what "Evolution Scientists" have to say about their superhero Mr. Charlie Darwin and his infamous writings. Darwin's theory in relation to fossils is a theory and nothing more.

"Paleontologists have paid an exorbitant price for Darwin's argument. We fancy ourselves as the only true students of life's history, yet to preserve our favored account of evolution by natural selection we view our data as *so bad* **that we almost never**

see the very process we profess to study." *(Evolutionist, Steven Jay Gould "The Panda's Thumb" (1882), pp. 181-182)*

Darwin tried hard to provide us with a comprehensive theory, and that is all that can be said in its favor. Macbeth says it well:

"It seems that the standards of the evolutionary theorists are relative or comparative rather than absolute. If such a theorist makes a suggestion that is better than other suggestions, or better than nothing, he feels that he has accomplished something even if his suggestion will obviously not hold water. He does *not* believe that he must meet any objective standards of logic, reason, or probability." - *Norman Macbeth, Darwin Retried (1971), pp. 71-78.*

His theories have been found to be inadequate, outmoded, and invalid.

"I assert only that the mechanism of evolution suggested by Charles Darwin has been found inadequate by the professionals, and that they have moved on to other views and problems. In brief, classical Darwinism is no longer considered valid by qualified biologists." - *N. Macbeth (Darwin Retried -1971)*

Darwin himself admitted that the evidence for evolution - which should be found in the fossil strata - simply was not there.

"Charles Darwin, himself the father of evolution in his later days, gradually became aware of the lack of real evidence for his evolutionary speculation and wrote: 'As by this theory, innumerable transitional forms must have existed, why do we not find them embedded in the crust of the Earth? Why is not all nature in confusion instead of being, as we see them, well-defined species?'" - *H. Enoch (Evolution or Creation (1968), p. 139)*

Knock Knock. "WHO'S THERE?" Darwin. "DARWIN WHO?"
That's what they'll be saying 50 years from now... Darwin WHO?

Darwinism is a belief in the "Meaninglessness of Existence." There's no hope for an eternal life with our true creator.

"Darwinism is a creed not only with scientists committed to document the all-purpose role of natural selection. It is a creed with masses of people who have at best a vague notion of the mechanism of evolution as proposed by Darwin, let alone as further complicated by his successors. Clearly, the appeal cannot be that of a scientific truth but of a philosophical belief which is not difficult to identify. Darwinism is a belief in the meaninglessness of existence." - *R. Kirk ("The Rediscovery of Creation," in National Review (May 27, 1983), p. 841)*

Darwin's underlying objective was to... FIGHT AGAINST GOD!!

"The origin of all diversity among living beings remains a mystery as totally unexplained as if the book of Mr. Darwin had never been written, for no theory unsupported by fact, however plausible it may appear, can be admitted in silence." - *L. Agassiz (On the Origin of Species, American Journal of Science 30 (1880), p. 154)*

Evolutionary theory may not be the root of the tree of evil, but it lies close to it. The root is the love of evil; evolution provides an excuse for continuing that indulgence.

"This monkey mythology of Darwin is the cause of permissiveness, promiscuity, pills, prophylactics, perversions, abortions, pornography, pollution, poisoning, and proliferation of crimes of all types." - *Braswell Dean (1981 statement, quoted in Asimov's Book of Science and Nature Quotations, p. 92 (Atlanta Judge)*

Denton, a careful Australian scientist, gets to the heart of the problem: There is no evidence for the theory.

"[Darwin's theory that all evolution is due to the gradual accumulation of small genetic changes] remains as unsubstantiated as it was one hundred and twenty years ago. The very success of the Darwinian mode at a microevolutionary level [finding change within species]... only serves to highlight its failure at a macroevolutionary level [finding change across species]." - *Michael Denton, Evolution; A Theory in Crisis (1985), pp. 344-345)*

While he was still alive, Darwin himself admitted his speculations were far beyond true science.

[In a letter written to Asa Gray, a Harvard professor of biology:] "I am quite conscious that my speculations run quite beyond the bounds of true science." - *Charles Darwin (quoted in N.C. Gillespie, Charles Darwin and the Problem of Creation (1918), p. 2 - University of Chicago book)*

It is all just a MYTH.

"Ultimately the Darwinian theory of evolution is no more nor less than the great cosmogenic myth of the twentieth century . . the origin of life and of new beings on Earth is still largely as enigmatic as when Darwin set sail on the Beagle." - *Michael Denton (Evolution: A Theory in Crisis (1985), p. 358)*

Fallacious solutions without any real answers.

"The theory of evolution gives no answer to the important problem of the origin of life and presents only fallacious solutions to the problem of the nature of evolutionary transformations." - *Jean Rostand (quoted in G. Salet, Hasard et Certitude: Le Tiansformisme devani la Biologie Actuelle (1973), p. 419) - Quotes above from EvolutionFacts.com*

BOTTOM LINE... **CHARLIE WAS WRONG...** DEAD WRONG!!

"The Spirit of God has made me, and the breath of the Almighty gives me life."

- Job 33:4

The Incredible Human Body
(Protein Molecules, Brain, Heart, Cells, DNA, Lungs, Eyes, Ears, Etc)

LIVING FORMS ARE TOO AWESOME TO REGULATE THE TENDER MERCIES OF **TIME** AND **CHANCE**. IT TOOK **SPECIAL DESIGN**, **SPECIAL THINKING**, AND **SPECIAL POWER** TO CREATE ALL LIVING BEINGS.

Almost all branches of modern science were founded, co-founded, or dramatically advanced by scientists who believed in the *biblical* **account of special creation.** In the US alone, it is conservatively estimated that there are upwards of 10,000 professional scientists *(the vast majority not officially linked to creation organizations)* who believe in *biblical creation. - Carl Wieland (Stone and Bones - 2011, pg 3)*

The *mathematical* **probabilities of the universe itself and all life on Earth happening by** *pointless, undirected chance* **(evolution) are** *Less Than Zero.* OK, maybe not less than zero, but you get my snarky point. Actually, a deep study of the different theories of evolution is great for a good laugh.

The Human Body

We live in an amazing world. The greatest of all of God's creations is man himself, the marvelous machine - precise and efficient. The human body has a dynamic framework of bone and cartilage called the *skeleton.* The human skeleton is flexible, with hinges and joints that were made to move. But to cut down harmful frictions, such moving parts must be lubricated.

Man-made machines are lubricated only by outside sources. Amazingly, the body lubricates itself by manufacturing a jelly-like substance in the right amount at every place it is needed. Yes, the body is a wonder machine, despite the defects from genetic copying errors (mutations) that have accumulated since the fall of man brought on the curse (Genesis 3).

The body has an extremely intricate *chemical plant* within it. This plant changes the food we eat into living tissue. It causes the growth of flesh, blood, bones and teeth. It even repairs the body when parts are damaged by accident or disease. Power, for work and play, comes from the food we eat. How in the world could this all have just *evolved* over millions of years, let alone with all of the different species on earth?

Even in freezing weather, our bodies will sometimes overheat. The body's own cooling system then takes over. Drops of perspiration pour from millions of tiny sweat glands in the skin. This is a major way in which our cooling system keeps our temperature down. **The human body has an automatic thermostat that takes care of both our heating and cooling systems, usually keeping body temperature at about 37°C (98.6°F).** Do you really think this just *evolved by random chance processes,* over long periods of time? *Another impossibility.*

The *brain* is the center of a complex computer system *far* more wonderful and complex than the greatest one ever built by man. **The body's computer system computes and sends throughout the body, billions of bits of information, information that controls every action, right down to the flicker of an eyelid.** In most computer systems, the information is carried by wires and electronic parts. In the body, nerves are the wires that carry the information back and forth from the central nervous system. And, in just one human brain there is probably more wiring, more electrical circuitry, than in all the computer systems of the world put together. Yes, it is a wonderful thing - this brain of ours.

In fact, as we look at this very moment, we are actually seeing with our brain. Although, of course, the message is carried there from another marvelous structure, the human eye. Modern cameras operate on the same basic principle as our eyes. In our eye the focus and aperture are adjusted automatically. Another marvel of God, *not* "lucky chance, zero odds, brain dead," evolution.

The sound we hear is being played on a perfect little musical instrument inside our ear. The sound waves go down the auditory canal and are carried by the bones of the middle ear to the cochlea, which is rolled up like a tiny sea shell. More on the ear in a bit.

The heart actually is a muscular pump forcing blood through thousands of miles of blood vessels. Blood carries food and oxygen to every part of the body. The heart pumps an average of six liters (1.5 U.S. gallons) of blood every minute, and in one day pumps enough blood to fill more than forty 200 liter (50-gallon) drums.

Yes, the human body is a wonderful machine. The fact that any one of these devices exists is a complete demonstration that they are the work of an *intelligent* and *skillful designer*, God Himself.

"God created man in His *own* image, in the image of God created He him, male and female created He them." - *Genesis 1:27*

The raw material, the basic chemicals in our body, can be found in the *dust of the ground*. However, these chemicals *cannot* arrange themselves into cell tissues, organs and systems. **This can *only* happen with an input of... *intelligence*. No chance it was *chance*.**

The book of Genesis teaches that God took 'the dust of the ground,' a heap of chemicals, shaped a man and then blew into his nostrils the breath of life. Then man became a living soul. Human beings are different from animals, for **"God created man in his *own* image."** - *Genesis 1:27.*

Our bodies have been *designed* with the ability to pass on to the next generation the programmed information required to form another person from simple chemicals.

We are more than the chemicals that form our body. *We are a special creation of God.* Man is God's masterpiece - His workmanship, the crown of creation. - *Creation.com*

Miracle of a Baby

Life begins at conception as a single, separate, living cell. Nothing new is added except oxygen and nutrition. Fertilization occurs when a sperm and ovum join to form a single cell, full of life and bearing the unique genetic imprint of a person who has never existed before. **The DNA in the 46 chromosomes of that small cell contain full instructions about that new individual's sex, eye color, foot size, brain capacity, and other physical traits. Mr. Darwin had *no* idea what DNA was back in his day. It turns his mighty theory into a Fairytale.**

Let's take a quick look at God's amazing work...

Week 1: Implantation... On about the sixth day, the growing baby attaches to the wall of the mother's uterus. That rich nutrient lining welcomes the tiny tenant, and soon the child is sending out the chemical signal that can be detected in a home pregnancy test. Before the second week is over, the cells of the child's body will have already begun segregating themselves into the various layers that will give rise to the brain, nervous system, skin, digestive system, muscles, bones, and circulatory system.

Weeks 3-4: A Beating Heart… The baby's heart begins its first beats as early as 18 days after fertilization, usually *before* the mother even suspects she is pregnant. Between the third and fourth weeks, the baby's head and spinal column become easily distinguishable, and arm buds appear. Legs will begin to appear days later. The umbilical cord forms, transporting oxygen and nutrients to the child.

Week 6: Brain Waves… Fingers are forming, and the child's mouth and lips are apparent. The child begins his or her first movements. At six weeks, the baby has brain waves that can be measured with an electroencephalogram.

Weeks 10-11: Organ Systems in Place… The baby has eyelids, fingernails, and fingerprints, and can grasp an object. The kidneys begin to form urine. All body systems are in place and active: the baby has a skeletal structure, nerves, and circulation.

Week 12: Movements and Characteristics… Though too small to be felt by the mother, the baby reaches peak frequency of movement during the third month. The baby's sex can be visually determined, and the child's eyes, ears, and face begin to display distinctive characteristics.

Week 14: A Miracle of Development… Eyebrows have formed, eye movements are seen. For two of weeks, this baby has had all the body parts required to experience pain, including the nerves and spinal cord.

Week 16: Making His or Her Presence Known… The baby becomes large and active enough for the mother to feel movement of turns, kicks, and somersaults that at some point even become visible to the outside.

Week 20: Hearing Mommy's Voice... In the fifth and sixth months, the baby responds to music, sudden noises, and voices, especially that of his or her mother. Over the coming weeks, the baby will increase seven times in weight and nearly double in height.

Week 23 - or earlier: Viability... Viability is the time when the baby can survive outside the mother. Not long ago, viability was at 30 weeks, then 25. Today, babies at 22 or 23 weeks have been saved, and even some younger babies have survived.   - *Crossway.org* (From the tract: *Life Before Birth*)

TIME TO THINK... So, do you still think you just *evolved* inside your mom's belly in 9 months(ish) by some *misguided, accidental, random chance processes of nature?* I would hope not.   The *mathematical odds* of that are *zero.*

For several months we're basically living in water inside another living human being, then instantly one day we're breathing oxygen in the *outside* world. If put back into water again for more than a few minutes, we'll die.   Pretty amazing, isn't it? We might call it *miraculous,* maybe because it's nothing short of a miracle!

A baby's body has about 300 bones at birth. These bones eventually fuse together to form 206 bones as we become adults. Another act of *intelligent design.* We could go on and on. There's a higher power at work in every birth, a *God who loves and adores YOU* and has a *purpose* for your life. His love letter to you, the Bible, talks all about that purpose and His love for you. If you take the time to read it, you'll be amazed at what He has in store for you. Take the time. Make the time. It's the greatest book you will ever read.

If you're a *non-creationist,* don't allow the *"theory of evolution"* to steal your life from you any longer. You can (and should) change that *"death" worldview* today!! God can change the hardest of hearts, but only if you ask Him to. He *won't* do it against your will. Please *don't wait,* you never know what tomorrow may bring.

"For you formed me in my inward parts; you covered me in my mother's womb. I will praise you for I am fearfully and wonderfully made." *- Psalm 139:13-14*

Amazing DNA *(Disproves Evolution in So Many Ways)*

DNA is... the carrier of the *inheritance code* in all living things. It is like a microscopic super computer with a built-in memory.

DNA stores a fantastic number of *blueprints,* and at the right time and place issues orders for distant parts of the body to build its cells and structures.

HERE'S A TOTALLY IGNORANT CLAIM... that LEADING *EVOLUTIONISTS* applied their theory to the amazing discoveries about DNA (discovered 100 years *after* Darwin's little book on origins) and... **came up with this Uneducated Claim (AKA: FAIRY TALE):**

All the complicated DNA in each and every life form made itself out of dirty water back in the beginning! You see, there was some gravel, along with some dirt. Nearby was some water, and overhead a lightning storm. *Presto-Chango...* The lightning hit the dirty water and made living creatures, complete with DNA. They not only had their complete genetic code, but they were also immediately able to eat, digest food, move about, perform

---

enzymatic and glandular functions, and all the rest.Instantly, they somehow automatically knew how to produce additional cells, their DNA began dividing *(cells must continually replenish themselves or the creature quickly dies)*, their cells began multiplying, and every new cell could immediately do the myriad of functions that the entire creature *must* do in order to survive longer than a couple *minutes.*

That same bolt of lightning made both a male and a female pair and their complete digestive, respiratory, and circulatory organs. It provided them with the complete ability to produce offspring and they, in turn, more offspring. That same lucky bolt of lightning also made their food, with all its own DNA, male and female pairs, etc., etc. I guess mating just came natural.

And that, according to this fun little children's story, is where we all came from boys and girls! It's a story that only very little children would find believable. Or is it? It's hard to believe that thinking adults actually believe all this twisted fairytale trash talk.

DNA is found in the cells of every living thing, except mature red blood cells. DNA is a complicated *language system* discovered around 1953.

Your own DNA is scattered all through your body in about *100 thousand billion* specks, which is the average number of living cells in a human adult. **The code within each DNA cell is *extremely complicated!*** If you were to translate all the coded DNA instructions from just ONE single human cell into English, it would fill many large books, each one the size of an unabridged dictionary!

DNA is *not* just chemistry; it holds all of the information needed to build a plant, a dog, a cat, or a human. Even Batman. Obviously, you must *know the language* to be able *to read the information.*

Most evolutionists say that monkey and human DNA are about 98% identical. First of all, it's actually been proven to be closer to 88% (If you look even deeper, it's far less than that because most DNA tests have been greatly skewed to show the results evolutionary scientists want and need to fit their anti-creation worldview. More can be learned from *Dr Georgia Purdom* at <u>AnswersInGenesis.org</u>). **It's said that** *mice* **and humans share 90% of the same DNA. Humans and** *chickens* **are 65% the same, and even** *bananas* **share 50% of the same DNA as humans.** Similar DNA simply means a *"common creator,"* certainly *not* evolution. **You can use any** *statistics* **you want, but I still ask you, "Where are the** *transitional fossils* **between species?" That line may be getting old, but they are the evidence you need to provide if you are to be taken seriously.**

Language *only* **ever comes from** *intelligence,* **and information** *only* **ever comes from** *information.* **You need** *both.* **DNA cries out loud and clear that there is an** *Intelligent Designer,* **NOT <u>non</u>-***intelligent accidental chance.*

DNA represents a language made up of 4 basic units called *nucleotides* that combine into a code for 20 amino acids that form over 100,000 proteins, that are the building blocks for every living thing. The similarities between DNA and human language are uncanny.

DNA represents the *language of life* **found in** *every living thing.* *Charles Darwin was not aware of DNA and the catastrophic problems it would cause his theory after his death.*

Evolutionists believe that *"matter"* **gave rise to life. However, first,** *matter* **had to produce a** *"language system,"* **then** *DNA,* **and then finally** *"life."* Matter had to give rise to multiple zillions of bits of information over millions of years – *new* information. This is *mathematically impossible* to the highest degree. Yet, another strong case for a *Supernatural God* and not the paltry theory of evolution.

Keep in mind that **nobody has ever seen** *matter* **produce ONE BIT of brand new information** apart from what *already* existed.

Richard Dawkins (well-known atheist kingpin) stated in the movie EXPELLED, that life on Earth was started by superior beings (aliens) from another galaxy/solar system that evolved to a higher degree. (To which I ask, "You know this HOW, Mr. Dawkins?").

How does our solar system keep its pants up?
With an asteroid belt. (HaHa)

For the evolutionary process to actually work, it *must* **add** *new* **information to living things. What we see, however, is a** *loss* **of information or information simply being** *rearranged.* (Can you say, *"Another nail in the evolution coffin"?* I'll bet you can).

Science proves **that the** *DNA code* **and the components of** *protein* **alone are so** *extremely* **complicated that it makes the possibility of** *any* **life being produced by** *haphazard chance events...* **ZERO.** Fact is, it could *never* thereafter evolve into new and different species! Each successive speciation change would *require highly exacting code* to be in place on the very first *day* of its existence as a unique new species. - *EvolutionFacts.com*

YES, A REAL "ADAM AND EVE"... Did ALL MANKIND really come from ONE MAN and ONE WOMAN? If you think that Adam wasn't a real man, explain the Biblical genealogy from him to Jesus... To add to that, recently scientists in the U.S. and Switzerland discovered that **DNA research now proves that ALL HUMANS came from** *ONE MAN* **and** *ONE WOMAN* **about SIX** *THOUSAND* **years ago.** Let's rightfully call them *Adam and Eve.* What makes this an even stronger case for *Intelligent Design* is that the researchers believe in an *evolutionary explanation* and are *not* making a case for a biblical *Adam and Eve*. Yet, it fits the creation model that the Bible has said from the very beginning.

Most evolutionary scientists have stated that 95% of DNA is useless, and call it **Junk DNA,** trying to discredit it as any evidence for intelligent design. This *false, uneducated* and *unsupported* claim has been completely and scientifically *disproven,* as top scientists have found that virtually *all* DNA has great value, meaning there's *NO such thing as - Junk DNA*. Many recent, credible studies back this up. So there you go Uncle Charlie.

Translation Package Needed at the Very Beginning

The amount of information in **the genetic code is so vast that it would be entirely** *impossible* **to put together by unguided** *chance,* **such as evolution.** But, in *addition,* **there** *must* **be a means of** *translating* it so the tissues can actually *use* the code.

BOOM EVOLUTION... **According to the theory of evolution, not only did the DNA have to originate itself by random accident, but the translation machinery** *already* **had to be produced (by accident), too - and also** *immediately!* **Without it, the information in the DNA could** *not* **be applied to the tissues. Instant death would be the result. Friends, evolution is oh so DEAD!!**

This translation package has also been termed an *"adapter function."* **Without a translator, the highly complex coding contained within the DNA molecule would be** *useless* **to the organism. This is something that mere chance could** *not* **possibly accomplish.**

Protein *(Yet Another Way of Making Evolution Impossible)*

One of the most important discoveries of the twentieth century was the discovery of the *DNA molecule.* It has had a powerful effect on biological research. **It has also brought quandary and confusion to evolution scientists. If they dared to admit the implications of DNA, it would also bring** *total destruction* **to their theory.**

Putting Protein and DNA together will not create life, but on the other hand, **there can be** *no* **life without BOTH Protein and DNA. Proteins would also have had to be made** *instantly,* **and in the** *right combination* **and** *quantity,* **at the** *very beginning.* Do not forget the sequence: **Protein** *must* **be in its proper sequence, just as DNA** *must* **be in its correct sequential pattern.**

Proteins come in their own complicated sequence! They have their own coding. That code is *spelled out* in a long, complicated string of materials. **Each of the hundreds of different proteins is, in turn, composed of still smaller units called** *amino acids.* There are twenty essential amino acids (plus two others not needed after adulthood in humans). The amino acids are an *extremely complex* assortment of *specifically arranged* chemicals.

Making those amino acids out of *nothing,* **and in the** *correct sequence,* **and doing it by** *unplanned chance* **would be just as** *mathematically impossible* **as a** *chance formation of DNA code!*

PROTEINS AND HYDROLYSIS - Even if protein had been made by chance from nearby chemicals in the ocean, **the water in the primitive oceans would have hydrolyzed (diluted and** *ruined***) the protein.** The chemicals that had combined to make protein would immediately reconnect with other nearby chemicals in the ocean water and **the protein would** *self-destruct!*

A research team at Barlian University in Israel, said that **this complication would make the successful formation of just one protein totally** *impossible,* **mathematically.** It would be 1 chance in 10 to the 157th power. **They concluded that** *no* **proteins were ever (or** *could* **ever be) produced by** *chance* **on this Earth.**

Another study said that the improbabilities of **Darwin's theory** start with the *protein* molecule. **The mathematical probability of the first proteins forming** *spontaneously* **by** *futile chance* **is less than 1 chance in 10**65 **power. That's like finding the winning lottery ticket on the street** *every week*, **for 1,000 years.** So, in other words it's completely **impossible!!** *- EvolutionFacts.com*

"There are 2000 complex enzymes required for a living organism, yet not a single *one* **of these could have been formed on Earth by shuffling processes in even 20** *billion* **years!"** *- Fred Hoyle (Famous Astronomer - November 19, 1981 issue of New Scientist).*

A Beating, Loving Heart

Are we so naive and gullible to believe that our hearts also evolved by *happenstance?* It just all *fell into place* somehow? Or, do we realize that it could *only* have happened by a loving God, the Creator of all, who puts His love in our amazing beating hearts that give us life.

Trust in the Lord with all your *heart* **and lean** *not* **on your own understanding.** *- Proverbs 3:5*

Heart tissue, like all cells in the body, needs to be supplied with oxygen, nutrients and a way of removing metabolic waste. This is achieved by the coronary circulation, which includes arteries, veins, and lymphatic vessels. The embryonic heart begins beating at around 22 days after conception. The heart functions as a pump in the circulatory system to provide a continuous flow of blood throughout the body. - *wikipedia.org*

The heart is almost pure muscle and pumps blood through *thousands of miles* of **blood vessels. If laid end to end they would *circle the Earth, twice.*** Blood carries *food* and *oxygen* to every part of the body as your heart pumps about 5 liters of blood through your body every minute.

This all *didn't* just happen by *accident* somehow. **A baby grows the heart and blood vessels in 9 months in its mommy's belly, so how can this possibly be if evolution says we took millions of years to develop?** It is interesting how people can accept evolution as real science, when on so many levels it makes *no* sense whatsoever. In fact, **it's beyond the scope of possibility. It's actually *anti-science.***

While we're asking questions that evolution has *no* answers for, let's ask a few more like... *where* did the *blood* come from in the first place? *How* did it know it was necessary? *How* did the blood vessels know to build and work with the heart? How did they know *where* to deliver all the oxygen and nutrition? And, with all our human wisdom, money and resources, *why* can't we create even *one* blood cell? **If we trust in the dirty, filthy *lies* of evolution,** it just makes us look uneducated and naive.

The Cell

Cells are the basic building blocks of all living things. **All living things are made up of cells.** The primary purpose of the cell is to *organize*. **Cells have *many* parts, each with a *different* function.** Cells were discovered by Robert Hooke in 1665. He named them because of their resemblance to cells lived in by Christian monks.

Here are a few questions that come to mind concerning cells... Since cells continue to regenerate throughout our lifetime, and it's said that our body replaces itself about every seven years or so, and it's estimated that there are between 37 *trillion* and 100 *trillion* cells in the human body (over 200 *different* kinds)... ***How* do cells know when to *stop* reproducing in our bodies so as not to produce too many and kill us?** ***How* and *why* do between *100 million and 300 million* cells *die* every minute, only to be *replaced*?** ***How* do all the different kinds of cells know how to work *together* to perform the specific functions they do?** Of course these are all rhetorical questions. HINT: The answer to them all is... *not* by *random chance processes, but rather by the Intelligent Design of a Supernatural Creator.* This is the *only* realistic and intelligent scientific choice.

The proven scientific *impossibility* of the incredibly complex *first cell* ever forming and actually staying alive, evolving and then reproducing is one of the most indisputable proofs *against* evolution, and literally crushes the ineffectual theory to smithereens. (As you can see, I love using fun old words). **Cells must continually *replenish* themselves or we quickly die.**

Believing we are all here by some *undirected accident* - that the first cell came from outer space or evolved out of a slimy mud pit, and that over time it evolved into a fully formed heart, brain, lungs, muscles, skin, eyes, central nervous system, etc. is only science *fiction*. Scientifically speaking, these are all insurmountable, hopeless tasks.

And, believing that all the *first* supposed evolved species (full of different cells) evolved over super long periods of time, into multiple thousands of completely *different* kinds of creatures, such as the first single cell evolving into *whatever, or* an acorn growing into a rose bush, or a fish evolving into a land animal, or a monkey into a human... are only *fairytales.* Lies, lies and more flat out lies.

Unless *all* of our essential body parts were in their *PROPER PLACE,* and *FULLY-FORMED* in short order (*not* over long periods of time)... LIFE COULD <u>NOT</u> EXIST. Period!

Even only *one* simple *protein* (inside a human cell) has the mathematical chance of 10 to the 50th power of forming by *accident,* as is the evolutionary way. In other words, it was *completely impossible* for you and me to evolve into the infinitely complex flesh and bone super machines we are, no matter if we had zillions of years to do it. By the way, **there are about 42 *million* protein molecules in *every* cell.** Think of these expanded mathematical probabilities. Not good for *evilution* is it? (Oops, I meant *evolution*).

"Within the period of human history we do *not* know of a single instance of the transformation of a single species into another one." - *Dr. T.H. Morgan (Nobel Prize winner and Evolutionist)*

Atoms

An *atom* is made up of three particles called subatomic particles: *protons, neutrons, and electrons.* The protons and neutrons make up the center of the atom, called the *nucleus.* The electrons orbit the nucleus.

Everything in the universe (except energy) is made of matter, thus, *everything* in the universe is made of *atoms*. An atom is a million times smaller than the thickest human hair. Two *hydrogen atoms* and one *oxygen atom* combine to make a *water molecule.*

A Google search tells us that around 450 B.C., the Greek philosopher Democritus introduced the idea of the atom. However, the idea was essentially forgotten for more than 2,000 years. In 1800, John Dalton re-introduced the atom. He provided evidence for atoms and developed atomic theory.

Again, it's estimated that there are up to **100 *TRILLION* "cells" in our body.** That's a big difference but either number is a whole lot of cells. **A single living cell may contain *one hundred thousand million "atoms."*** Talk about tiny. **Each and every atom is arranged in a *specific* order.**

Atoms are based on exact *design* and *design requires intelligence.* So, random chance evolution *ain't* gonna cut it.

Let's go back to the very first moment that the supposed *Big Bang* took place. Evolutionists say that at that very first moment, everything was left to *chance.*

Now let's move forward to what we actually know about the various *atoms* and their structure. **Life as we know it is based on the *carbon atom.*** Its unique structure makes it the only atom with almost unlimited ability to share pairs of electrons with other atoms. This makes possible the rich range of biological molecules needed for life. *No* other atom can do carbon's job.

The *oxygen atom's* structure causes it to bind together in pairs. This type of bonding leaves unpaired electrons that allow oxygen to bind with iron. This feature makes hemoglobin capable of carrying oxygen in the blood. There are several other atoms that could replace iron in hemoglobin, but they would hold the oxygen either too tightly or too loosely. So there are *no* substitutes for iron.

Likewise, the *zinc atom* is the only atom that can allow proteins to do the crucial job of identifying their own unique DNA sites.

The precise *structure of atoms* was clearly *not* the result of unorganized chance. Each was carefully designed by the Creator to support the life He would form only days after He made the atoms.
- *Paul A. Bartz, CreationMoments.com (with additions from the author of this book)*

Gills to Lungs *(Sounds Fishy to Me)*

Evolutionists *can't* scientifically explain the evolution of fish gills to land animal lungs, or fins to legs and arms. This is in addition to many *other* body parts we won't even attempt to discuss here. Let your imagination run wild. Well, at least the *evidence* is plain to see. Oh yeah, there is NONE!

They simply claim it happened, so I guess the rest of us are just supposed to go with the program. Not me, and hopefully not you either. God gave us a brain to figure nonsense like this out. We need to call the truth stretchers out on the carpet.

So, let me get this right… one day fish can't live on land for more than a few minutes before dying. Then magically, somehow they can live on land and have developed *lungs* to breathe, *legs* to walk, and *all the other body parts* they need to survive more than a New York minute? Great *science fiction* story, but that's about it. Again, NO EVIDENCE!

Yet, there are currently over 33,000 known extant *(living today)* fish still in the waters of the world who *didn't* and *don't* evolve. Huh? The evolutionary *fairytale for all ages* continues.

Despite the fact that yet again "NO" *Transitional Fossils exist* **anywhere** (and it would *have* to be quite substantial, *not* just some *cartoon illustrations* in high school or college science textbooks like we see now. Funny how people actually buy into this fish tale; hook, line and sinker. OK, enough of the fishy references)... **how in the world would they live more than a few** *minutes* **if it took millions of years to evolve these completely different (and** *essential for life)* **body parts?** They wouldn't. They couldn't. It's not only scientifically impossible, but it's just common sense. Think about it for a second and a half, and *fish to land animal evolution* starts to look pretty fishy.

"Evolution became in a sense a scientific religion; almost all scientists have accepted it and many are prepared to *bend their* *observations* **to fit in with it."** *- H. Lipson (Evolutionist. "Physicist Looks at Evolution" Physics Bulletin 31 - 1980, p. 138)*

The Eyes Have It

What is it that the eye's have? They have a *wonderful design.* Evolutionists like to claim the eye is evidence of evolution due to it's *bad* design, thus it must have evolved, as an intelligent Creator would not have made it this way. That's about how deep the argument on eye evolution really gets.

This is so ridiculous. "So, *where in the fossil record,* **Mr. Evolutionist, is the supported** *evidence* **that all animals started with** *no eyes,* **then** *10% little slit eyes,* **then** *half eyes* **and** *finally* *complete eyes."* **All we see are** *full eyes* **in virtually all of the creatures in the fossil record.** Creationists don't need to even begin to entertain these *stupid assumptions* stated in the Darwinian theory of evolution! Don't feel obligated.

The eye is a chemical and technological wonder. It can differentiate between about 10 million colors. Just the retina alone has 150 million *properly designed* and *placed* cells.

It would take thousands upon thousands of *carefully directed* **mutations for an eyeless creature to develop sight.** Since a partial eye doesn't provide sight, each of these thousands of *unguided mutations* would offer absolutely *no* advantage to the creature. So, from an evolutionist's view point... **WHY would this happen, and HOW would it know it was a positive change each time one needed to be made?**

Why and How did *any body part or system* **know it needed to first** *start* **and then** *continually evolve?* Why indeed.

The fact is... mutations *don't* **happen this way. All the true mutations we know about have** *harmed,* **rather than helped the creature with the mutation.** This fact places the thought of thousands of positive mutations in a row outside the realm of scientifically reasonable consideration.

Neither animals living today, nor known fossil animals, show any *evidence* **of gradual development of eyes. Therefore, the only reasonable** *scientific* **conclusion is that there's a** *Master Creator* **who designed and created the eye.**

The lens of the eye is a marvel of chemistry. It is made up of a concentration of protein molecules inside water-lined cells. When scientists learned this, they were amazed. Protein molecules in water are *not* transparent, as the lens must be.

The high concentration of protein molecules in the lens of the eye causes the proteins to pack together something like the molecules of window glass. As a result, **the normally opaque protein solution in the lens becomes transparent. This is God's amazing little secret.** The eye allows us to see our Creator's fingerprints all around us!
- *CreationMoments.com*

So, if the eye took millions of years to evolve, as evolutionists claim, then *how* and *why,* all of a sudden, do ours evolve in just *weeks* in our mother's belly before we're born?

Charles Darwin was frustrated by the eye's complexity, even though he knew only a fraction of what scientists have now discovered about the eye. He was *right* in doubting that an eye could evolve, so I guess maybe **Charlie WASN'T Wrong** all the time after all. ;-)

"To suppose that the *eye* **could have been formed by natural selection, seems, I confess, *absurd in the highest degree."* -** *Charles Darwin (The Origin of Species – 1859, 6th edition, 1872)* Note: He did go on to defend evolution, but again, as usual, with no real evidence, or the knowledge of modern science, that was not available in his day - that *proves* that something as complex as the eye could never possibly evolve.

Indeed, the EYES have it. God's incredible design that is.

Our Waste Management Systems

Let me ask you a simple question that should **prove to you why our** *internal plumbing systems* **could** *never have evolved* **and** *must* **be** *fully-formed* **in our bodies from the very** *start.*

So, before our internal plumbing (waste management system) was constructed (evolved), our waste went... *WHERE?* It would have spilled out *inside* our bodies, that's where. A good thing? A healthy thing? Of course not. Rather, **if *our plumbing* was not set up correctly from the start, we would *all* die quickly. There would be NO life on Earth today.**

Maybe that doesn't sound all that *scientific* to you. It actually is. **Do we really even need to go beyond *Common Sense* here?** As with much of creation vs. evolution, I think not.

The Bigger Question is... did our waste management systems, along with our other life supporting internal body parts and systems that all work *together* making us the incredible creatures we are... actually *evolve* ever so slowly over long periods of time? You know the answer. The evolution of mankind theory is just so stupid!! It really is. Think about this the next time you relieve yourself.

The Brain

"Set your *minds* on things that are above, not on things that are on Earth." - *Col. 3:2*

The human central nervous system (CNS) consists of the brain and spinal cord, including roughly 100 billion nerve cells.

The major component of the brain is the *neuron*, which is basically a device for sending information from one place to another. **The brain has 10 billion neurons in it. Each *neuron* is like a *small computer*."**

The 3-D world we experience, with sights, sounds, tastes, smells, and bodily sensations, is essentially constructed by our brains. *God is so, so cool.* Evolution is so, so... *nothing!!*

"The human brain is the *most complex* and orderly arrangement of matter in the universe." *- Isaac Asimov*

The original signal is interpreted so that a person sees the world in vivid color, hears the sound of a waterfall, experiences the heat of a hot summer's day, feels a breeze on the skin, and smells the brats on the grill as they're getting ready to watch the *Green Bay Packers* play. Go Pack Go! This can happen all at the *same time,* really fast.

How does the brain do this? No one knows, but it must involve signal processing way beyond our current understanding. Evolution? *Gimme a break.* Sorry, I love using old school terminology sometimes, just for the fun of it. Actually, people still use "gimme" today in texting.

Both computers and brains will malfunction if physically damaged. However, the brain, depending on the nature of the damage, often has enough built-in redundancy and neuroplasticity (the ability to reorganize its connections) that other parts of the brain can take over the role of the damaged regions.

We are clearly *not* mindless, accidental robots whose actions are determined by our genes, or a chain of chemistry since the *so-called* Big Bang. *- Creation.com (Get the Creation magazine, it's the best. It won't have my comments added either. Bully for you)*

"For although they knew God, they neither glorified him as God nor gave thanks to him, but their *thinking* became futile and their foolish hearts were darkened." *- Romans 1:21*

Ears

Every sound we hear is being played on a perfect little musical *instrument* that was carefully created inside our ears, by *the Master Designer* who also created your ears.

The sound waves go down the auditory canal and are carried by the bones of the middle ear to the cochlea, which is rolled up like a tiny sea shell. The outer ear operates in air. But **the cochlea is filled with *liquid*, and transferring sound waves from air to liquid is one of the most *difficult* problems known to science. Yet, our ears do it.**

Three tiny bones called the ossicles are just right to do the job that enables us to hear properly. Interestingly, the size of these little bones *does not change* from the time we are born. - *Creation.com*

The very idea that such complex physiological mechanisms as our ears with the intricate nervous system that conveys their messages, and the magnificently organized brain that receives and uses them, could develop by *natural processes*, could only arise in hearts in desperate rebellion against God.

Belief in evolution is nothing but rebellious folly, yet evolution is taught as sober fact in schools and colleges everywhere. **No one has ever seen it happen, and no one knows how it could work. It is nothing but a feeble attempt to get away from God, for God, and *only* God can make a hearing ear.** - *ICR.org*

I could go into more detail but I think you **HEAR** what I'm saying.

Vestiges

Are there remnants of evolution in your body? The Darwinists say there are, and that they prove evolution is true. These are said to be *unneeded organs,* which your *animal ancestors* used and then passed on to you. Obviously, the proof is that you have useless, no longer needed, organs which are *vestiges* (left-overs) from your evolutionary ancestors. *Or so they say.*

School textbooks as recent as the 1960s listed over 200 vestigial (useless) structures in the human body, including the important *thyroid* and *pituitary glands!*

To date, not one dedicated evolutionist has been willing to have all (or even half) of their "vestigial organs" *removed.* To do so, would require taking out most of their endocrine (hormonal) glands!

The truth is that *the theory of useless organs* as *evidence* of evolution was based on rank 19th-century ignorance of those organs! In fact, **today ALL organs formerly classed as vestigial are known to have a specific and** *purposeful function* **during the life of the organism!** Unfortunately, as expected and true to form, that fact is *not* mentioned anywhere in the school textbooks. How odd. ;-)

Here's an example of just one of the *so-called vestiges…* **The Coccyx.** Another organ declared useless, by evolutionists, is the *coccygeal* vertebrae (the coccyx). This is the bottom of your spine, your *tailbone.*

Scientists have found that *important muscles* (the *levator ani* and *coccygeus) attach* to those bones. Without those muscles, **your pelvic organs would** *collapse; that is, fall down.* **Without them you could** *not* **have a bowel movement,** *nor* **could you walk or sit upright.**

No capable biologist today claims that *any* vestigial organs exist in human beings. - *EvolutionFacts.com*

Questions to Ask Yourself

WHY does my body have *circulatory, digestive, endocrine, immune, lymphatic, nervous, muscular, reproductive, skeletal, respiratory, urinary, and integumentary* **Systems**? **HOW** did they *form* and **HOW** do they all work *together* so well, or even work at all? **WHY** do I have all the *"essential for life"* parts in my body that I have; such as a *brain, heart, kidneys, liver and lungs?* And, **HOW** in the world did these also all *form* and work *together* so you and I could stay alive all these years? HOW and WHY indeed.

And, from EvolutionFacts.com: **Why do I have** *odor-detecting* cells in my nose? **Why** can I *taste* with my tongue? **Why** does food have *built-in flavors?* **Why** do I have *semicircular canals* in my ears *sending signals to my brain* so I can stand without falling over?

If evolution is true, we need to ask not just HOW, but also WHY would these things evolve?

Random (Created) Stuff
(Water, Oxygen, Carbon Dioxide, Gravity, Fire, Ozone, Memory, Food, Eugenics, Love, Etc)

THE MATHEMATICAL ODDS... of things like Oxygen, Carbon Dioxide, Gravity, Ozone, Plants, Water, etc. all *just perfectly falling into place* **to work together to sustain life on planet Earth, all happening by** *pointless, random chance happenings* **again and again a bazillion times...** is *zero,* **baby. As in a big fat ZERO. It's rather comical to think they could evolve, actually.** Wake up, and step into reality my evolution believing friends. The dream is over. The fact is... *Evolution is a Fairytale for All Ages* and many evolutionists are starting to doubt it more and more. Good for them.

Mathematical Probabilities

Do "not" just accept the shameless evolutionary theories about everything being created, living and non-living. For even any "one" of the many creation related topics talked about in this book to happen by *random chance,* the probability is virtually *zero.* When you add them all up, plus thousands more that are not in this book, you see that **creation could** *never* **happen without an** *All Powerful CREATOR* (who is more intelligent than His creation, and *not* the other way around where *un-intelligent matter evolved into intelligent creation*).

So, if you choose to believe all the evolutionary mumbo-jumbo, that it all just *happened/happens by uncontrolled CHANCE* **and that there was/is no grand** *intelligent design* **by an all powerful**

Creator of the Universe... MAYBE YOU NEED TO ASK YOURSELF... "WHY do I believe this? Could this all be a bunch of lies?" Could there really be a *Divine Master Designer?* If so, *how* can I get to *know* and *understand* this Master Creator?" The answer is that you CAN get to know Him, on a *personal* level. There are no costly seminars to attend, no extra prep-work involved, no promotions needed, no brown-nosing necessary. It's the GREATEST GIFT EVER and it's free of any financial cost.

Humans and Plants
(Best Friends Forever)

I guess you could say that humans and plants are *Best Friends Forever...* The perfect symbiotic relationship. Think about it, **besides all the essential life sustaining things like sunlight, soil, nitrogen, pollinators, and water to live - plants also require the *carbon dioxide* that "we" breathe out.** In return, we need the oxygen they release. Thus, **we** *need* **each other to live.** Coincidence? Random chance? No chance my friends.

This is NOT the type of accident that happens by *fortuitous* *chance!* Never, ever in a billion years. That's nonsense. **It's an impossibility for evolution and** *you* **know it.** This could *only* be possible through *intelligent design.*

Humans and plants would have to live in the *same geographic location* **at the** *same time* **in history, and have all their complex parts** *already* **in place and fully-formed before their mutual dependence upon each other could be established.** Read about *Irreducible Complexity* in this book to see why this makes sense and denies any chance for the possibility of evolution to be true.

Let me see, humans aside, if plants didn't have "any" of the other life givers they require (sun, soil, nitrogen, pollinators and water), they would not be here today. Still denying there's a God who created all this? Come on, REALLY? Everything just... *happened??*

These facts alone expunge the unabashed, bush-league, out to lunch, foolhardy theory of evolution. Let's continue to explore more so you can be even more positive, without any doubt, that evolution is only a *fairytale* for all ages.

Herbs for Healing

Whether you pronounce them *erbs* or *herbs,* herbs are all around us. We use them when cooking, and we use them for healing a variety of things that ail us. An herb is a plant or plant part used for its *flavor, scent,* or *therapeutic properties.* They are usually less expensive than conventional medicine. Many times they help our body heal itself.

What do you think the chances are of all the world's herbs just appearing by *random chance?* The fact that most are in someway beneficial to us, for a variety of reasons, is mind blowing. The only realistic answer is that they were all *intelligently designed* for our benefit, by a great God who loves and cares for us, His creation.

World's Oldest Tree

The oldest known tree in the world is the Methuselah tree in southern California. Dating trees is a inexact science. You can get all sorts of answers. The textbooks state that the oldest tree is 4,300 - 4,400 years old. Now that is interesting. If the world is millions of years old, why don't we have a tree somewhere that is 20,000 years old? Is it just a

coincidence that the oldest tree is 4,300 or so years old? I have a theory on that. I believe that about 6,000 years ago God created Heaven and Earth. About 4,300+ years ago there was a flood that destroyed the world. Therefore, to find a tree that is 4,300 years old fits fine with my theory. If you believe in evolution, it creates a little bit of a problem. If Earth is millions of years old, show me an older tree.
- *Dr Kent Hovind (Creation Science Evangelism Seminar 1999)*

If evolution were true, we should see trees that are tens and hundreds of thousands, even millions of years old. We don't.

Oxygen *(Need to Breathe)*

Did you ever stop to *think* about things like... HOW did oxygen come to be? HOW did oxygen *know* it was needed for life? HOW do plants produce oxygen? HOW did it all happen anyway? And, WHY?

To some these may sound like silly questions – but are they? Or, are they the questions most people never think about, but should be asking?

Why **is the Earth's atmosphere always around 21% oxygen, which just happens to be the proper amount we need to stay alive?** How does it just *stay* at this level? And, how is it that we need around 21% oxygen to live, and not 5%, 39%, 72%, or 100%? If the oxygen on Earth were 35% we'd explode, if it were 10% we'd suffocate to death.

These are valid questions we should be asking. If evolution is true, there *must* be *valid* scientific answers. Evolutionists can't just *explain everything away* and *change the focus of the conversation*. Answers to these difficult questions are mandatory.

If *plants* (who provide us with oxygen and food) **- and** *animals* (who supply plants with carbon dioxide) **didn't evolve at the** *same* **TIME in history, and in the** *same* **geographical LOCATION, how did** *any living thing survive?*

Obviously, evolution is not supported by facts. To believe otherwise will *not* work either *scientifically* or *mathematically.* You might as well throw *common sense* out the window, as well.

ATMOSPHERE WITHOUT OXYGEN... Could a *non-oxygen* atmosphere ever have existed on Planet Earth? It surely seems like an *impossibility,* and there is *no evidence* for it whatsoever, yet **evolutionary theorists** *(guessers)* **have decided that the primitive environment had to have a** *reducing atmosphere,* **that is, one** *without* **any oxygen.**

Now, the theorists do *not* really want such a situation, but they know that **it would be** *totally impossible* **(without** *supernatural* **intervention from an intelligent Creator) for the chemical compounds needed for life to be produced outside in the open air. If oxygen was present; then amino acids, protein, etc, could** *not* **have been formed. Thus, life could** *NEVER* **exist.**

So, in desperation, they have decided that at some earlier time in Earth's history, there was *no* oxygen - anywhere in the world! Then later, it *somehow* arrived on Planet Earth to save the day! Gee, this sounds kind of like a child's bedtime story. I'm getting sleepy.

There is *no evidence anywhere in nature* **that our planet** *ever* **had a** *non-oxygen* **atmosphere!** And, there is *no* theory that can explain how it could begin with a reducing *(non-oxygen)* atmosphere - which later transformed itself into an *oxidizing* one!

A reducing (non-oxygen) atmosphere never existed earlier on our planet. Yet, without it, biological chemicals could *not* form. If a reducing atmosphere had existed, so biological chemicals could form (and if they could somehow be injected with *life*), they would *immediately die from lack of oxygen!* This is an oxymoron if there ever was one.

NO OZONE LAYER... If there were no oxygen in the atmosphere, there would be *no* ozone either. **Without the ozone layer, ultraviolet light would *destroy* whatever life was formed, within minutes to hours.**

All this did *not* happen *randomly*. It happened on *purpose* through INTELLIGENT DESIGN. Something (*anything* for that matter) can *NOT* come from Nothing!! Ever!! *Randomness* **is neither true, nor scientific, and that's what evolution is supposed to be.** If randomness can produce living wonders such as those surrounding us, then highly intelligent scientists working in well-equipped, multi-million dollar laboratories, surely ought to be able to produce eyes, ears, and entirely new species in a few months or years. They have *never* been able to create even one tiny, itty-bitty blood cell.

Do you see why *evolution is such a mindless and deceptive theory* for so many reasons? It is very destructive and must be exposed for what it is... *EVILUTION.* (You'll find more on this and many others topics at - *EvolutionFacts.com and CharlieWasWrong.com*)

Carbon Dioxide

For a number of reasons, carbon dioxide is one of the most important gases on Earth. Plants use carbon dioxide to produce carbohydrates (sugars and starches) in the process known as photosynthesis. (In photosynthesis, plants make use of light to break down chemical compounds and produce energy.) Since humans and all other animals depend on plants for their food, photosynthesis is *necessary* for the survival of all life on Earth.

Carbon dioxide in the atmosphere is also important because it captures heat radiated from Earth's surface. That heat keeps the planet warm enough for plant and animal (including human) life to survive.
- *ScienceClarified.com*

Some Things to Ponder... How long would animals live without *oxygen* to breathe? And, **how long would plants live without the** ***carbon dioxide*** **we breathe out? Not long at all.** Without it, they could *not* make chlorophyll. **When plants** *take-in* **carbon dioxide, they** *give out* **oxygen. But a** *reducing atmosphere* **has neither oxygen** *nor* **carbon dioxide. Therefore** *no* **plants could either live or be available for food.**

Molecules to Man

Molecules to Man. That's what evolutionists claim happened in the beginning, but do you really buy that? Come on now, *really?*

The universe is made up of atoms, each containing protons, neutrons, and electrons. A *molecule* is an electrically neutral group of two or more *atoms* held together by chemical bonds. Examples of molecules are: H_2O (water), N_2 (nitrogen) and O_3 (ozone). If molecules contain atoms of different elements, that is known as a *compound.*

Science proves that the probability that natural molecules could have formed *spontaneously* **in a gaseous environment is virtually** ***ZERO.***

Many molecules necessary for life, such as DNA, RNA, and proteins, are so *incredibly complex* that statements claiming that they have evolved are absurd to the highest degree. Of course, it's not surprising that evolutionary claims lack any experimental support.

There is *no* reason (or evidence) whatsoever to believe that mutations or any natural process could *ever* produce any new organs - especially those as complex as the eye, ear, heart, lungs, or brain. - *CreationScience.com*

"All of these little pictures of molecules coming together to form the first cell are *fallacious and ridiculous*. The Origin of Life community has *not* been honest." - *Dr James Tour*

Memories

According to the dualistic interpretation of the mind, the self uses the brain, but the brain is distinct from the self. **Evidence from a new study shows that memory circuits assist the mind in recalling thoughts by associating them with sensations. The mind can use the brain's storage mechanisms to sort more important or urgent memories for faster recall.**

In *Scientific Reports*, David Stawarczyk and Arnaud D'Argembeau present the question they wanted to investigate:

Daily life situations often require people to **remember internal mentation** (i.e. mental activity), such as their future plans or interpretations of events. Little is known, however, about the **principles that govern memory for thoughts** experienced during real-world events. In particular, it remains unknown whether factors that structure the retrieval of external stimuli also apply to *thought recall,* and **whether some *thought* features affect their accessibility** in memory.

The conscious self uses memory as a tool. The *mind uses the brain; it is not a passive illusion conjured up by the brain.* The brain is active and ready, storing each sensation as we walk through daily life, but thoughts that involve *planning* are more readily recalled. It's as if the self tells the brain, "Remember this," and the brain obliges like a computer operator or librarian, pigeonholing the data where it can later be recalled more easily.

Choosing to Remember... Everyone has had the experience of intuition at the wrong time. You're busy with something else, when an idea flashes into your mind, perhaps a melody for a song you want to write, a project you want to start, or something you want to say to counsel a friend or family member. It's impossible to stop and write it down at the moment, but you want to remember it later.

Research on the stream of thought in daily life has shown that the mean duration of a thought segment is about 14 seconds, suggesting that people may experience between 4,000 and 5,000 thoughts during a 16-hour day.

Many of these thoughts are likely forgotten, the authors say, probably because there is only so much a person can focus on in the sea of perceptions and stimuli going on around us. **If the intuition is valuable enough to the self, the brain will assist future recall of the intuition, but is not causing it; the "user" or self asks the brain to store that thought in its physical hardware.**

To help the user recall what it has selected as important, the brain forms an association between the intuition and the external stimuli occurring at that moment. It might be a mental picture of what street you were driving on, what book you were reading, or what child was up at bat on the ball field when the thought occurred.

Thoughts of Evolution... The word "adaptive" in that first sentence may trigger thoughts of evolution, but **actually, many of the other words indicate** *intelligent design:* **information, decisions, future orientation, planning. This is what intelligent minds engage in.**

They are *opposite* **to the predictable outcomes of natural laws.** Gravity says an arm left to itself will fall. A mind says, "I choose to raise my arm." **A mind plans, decides, and looks ahead to potential realities that may be entirely unique and creative. Even while daydreaming, creative minds can be thinking of things "unrelated to their ongoing activity and decoupled from current sensory input."** A cheerful factory worker can be whistling as he works, thinking of the show it came from, or even adding his own lyrics to it, until interrupted by an event boundary - the quitting-time bell.

All the brain does is assist recall of such creative mental activities. The worker might recall the lyrics he was imagining by associating it with hearing the bell. The researcher might recall the novel experiment she was thinking of by associating it with getting on the turnpike. External cues and event segments help organize our thoughts. **The mind uses the brain; the brain is** *not* **the mind.** - *EvolutionNews.org*

Our minds have the capability of bringing up *past* **memories, even though some I'd prefer to forget. How about you? So, this just happens by** *random accident* **each time? Huh?? I don't remember how evolution explains this away but personally I'll give God the credit for our incredibly complex minds and memories. If you don't agree, please** *remind* **me again why I'm wrong.**

I forgot what I was going to write next. Apparently my memory is beginning to *devolve.* :-) Oh yeah, now I remember...

The Ozone Layer

The ozone layer *(or ozone shield)* **is a region of Earth's** *stratosphere* **that absorbs the majority of the Sun's ultraviolet radiation.** It's located about 9 to 18 miles above the Earth's surface, and just keeps hanging out there to protect life on Planet Earth. It's made by a *chemical reaction* where several oxygen molecules react with each other, however, this reaction takes place *only* when there's sunlight. In other words, as long as the sun is shining, ozone will constantly be produced. Pretty amazing if you think about it.

No Oxygen = No Ozone... Each ozone molecule has 3 oxygen atoms, unlike normal molecular oxygen. If there were *no oxygen* in the atmosphere, there would be *no ozone* either. **Without the ozone layer, ultraviolet light would** *destroy* **life on Earth within** *minutes to hours.*

Let's ask evolutionary scientists a few simple questions... *HOW did the ozone layer happen? WHY did it happen?* (I know they don't like the *WHY questions* as that suggests an *Intelligent Designer*). *WHEN did it happen? HOW* **does it** *continue* **to work?** Yes, these are actually valid questions that they don't have proper answers to.

I guess we're all so *lucky* **that the ozone layer knew it had to** *quickly evolve* (just as they say oxygen did) **to protect all life on Earth.** *Let the evolutionary fairytale continue...*

Many evolution scientists think us creationists are all a bunch of simpletons, and so lucky we have them to tell us *we're no more* than *molecules in motion* and living life *with false hope* because this life is all we get. Don't buy these trash talk lies. I'm so glad to be a creationist and understand *real* creation science. Aren't you too, my creationist friends? If you are, then share real creation news and facts with others. **Don't hold back and stay silent while evolution believers spread *false* rumors** about creation just to build upon the fairytales of evolution. **It's time we stand strong together for our Creator God and stop allowing the theory of evolution to destroy the lives of our family and friends.** I'm sure that sounds so dramatic to many evolutionists. It's not. It's truth.

God's Miraculous Water

WATER! You gotta love it. We drink it, wash in it, cook with it, swim in it and generally take it for granted. This clear, tasteless and odorless liquid is so much a part of our lives that we hardly ever think about its amazing properties.

We would die in a few days without water. Our adult bodies are made up of about 65% water. Water is necessary to dissolve essential minerals and oxygen, flush our bodies of waste products, and transport nutrients around the body where needed. **Water is the *only* substance on Earth that has these properties.** I'm getting thirsty just talking about it.

Did you ever think why **the Earth is the *only* place in the universe known to have *liquid* water**? And, this depends on having the right kind of star (our sun) neither too bright nor too dim, and thus neither too big nor too small. And, the planet must be at the right distance from it. *- Creation.com*

So, are we to believe that good 'ol H2O *evolved* from some *random chemical reactions* way back billions of years ago? And, this happened *how?* And, *why?* And, that *all* creatures and plants need water to live?

The probabilities of *hydrogen* and *oxygen* atoms actually forming from *nothing* is utterly ridiculous in the first place. Then the three (2 hydrogen molecules and 1 oxygen) *randomly hooking up* one day to create "water" are not even thinkable.

Water alone is just one more reason to know that the sketchy, bungling, atheistic theory of *chemical evolution* is 100% fictitious.

Seriously friends, use your (75% water filled) brain to understand that water could *only* come from God. It did *not* and could *not* magically appear one day out of thin air in any other way. Period!!

Water Tip: There are *two* reasons why you should never drink toilet water. (Number one. And, number two) :-)

Fire *(Only on Earth)*

Fire is one of mankind's greatest tools, yet it's also one of our most destructive forces. **Earth is the *only* known planet where fire can burn.** There is not enough *oxygen* on *any* other planet for fire to burn. *Random chance?* Fat chance.

Seed to Full Blown Plant

It's estimated that there are *almost 400,000 species of plants* on planet Earth. All starting from a wonderful little *seed*. All growing up to be a *different* plant in its own unique way.

Every SEED is a Miniature Miracle... and *always* produces after its *own* kind. Thus, an acorn seed will *always* produce an oak tree, but *never* an apple tree, rose bush, or spider monkey for that matter. All joking aside, **the inept theory of atheistic evolution claims that *every living thing is related* to each other, thus, all** *plants and* *animals are distant cousins,* if you will. **Note: Seed fossils** (dated many millions of years old by evolutionists) *look exactly the same* as **seeds we find today. This would *not* be so if evolution were true.**

Where did all the *"first"* seeds come from? How does an evolutionist begin to answer this question, not just for one kind of seed but for *all* the 400K *different* kinds of seeds since they *only* produce their *own kind?* From a scientific, mathematical, and common sense perspective, *being created by a supernatural God makes much more sense.* Evolution makes absolutely *no* sense whatsoever.

And, as far as the plant fossil record, as with animals, there are *no* transitional fossils to support a ho-hum evolution theory.

The actual plant lives in *miniature form* inside every seed. Keep in mind, **for a plant to survive, everything else would have to be in place *already* - like the *proper amount* of *carbon dioxide, nitrogen, phosphorus and other important nutrients, and of course* the *sun* and *water.* Oh yeah, we can't forget the *dirt.*** And, how did the roots know to grow *down* into the ground, and the plant to grow *up?* Seeds are programmed to stay dormant until water and warmth are available. How did this all happen? It sure *wasn't* pure chance. Nope, *not* evolution. No way, no how.

Another thing to think about… **among *LIVING* plants today we never see anything like a tree with some apples *and* some peaches.** Or a plant with some tomatoes and some beans. The buried *fossil record* also proves the same thing. **Each is very *unique* as our awesome GOD *designed* them. And, so are YOU.**

THINK ABOUT ALL THIS STUFF. Evolution is only a *fairytale.* God is your answer, *not* Uncle Charlie.

Flowers

"The more I study nature, the more I stand amazed at the work of the creator." - *Louis Pasteur*

The principal process for plant reproduction is *pollination*. The male organs of the flower, or anthers - produce pollen cells, or grains. These have to find their way to the female stamen, and then into the ovary, which contains the ovule cells.

Generally speaking, pollen does *not* pollinate the ovules in its own flower. So the plant has to have a strategy to get pollen from one flower to the ovary of another flower of the same species. To do this, it employs go-between pollinators.

The reason flower petals are so pretty is to attract insects to act as pollinators. So many flowers produce nectar – a sweet liquid that is desirable for many insects, especially bees and butterflies. Visiting insects will brush pollen off the anthers, and, if they already have pollen from another flower, they might brush that pollen on to the flower's stamen. But how do the insects know the nectar is there? By the bright petals that the flower produces!

These features above could *not* **evolve by** *irrelevant, random chance...* Brightly colored petals deprive a plant of energy. What would be the advantage of that if there were no pollinators? As we think about pollination – especially that of the more unusual plants, we realize that this is a process *designed by God.*
- *CreationMoments.com*

"And the Earth brought forth grass, and herb yielding seed after his kind, and the tree yielding fruit, whose seed was in itself, after his kind, and God saw that it was good." - *Gen. 1:12*

Hydrogen in the Universe

According to one theory of solar energy, *hydrogen is constantly being converted into helium* as stars shine. But hydrogen cannot be made by converting other elements into it. Fred Hoyle, a leading astronomer, maintains that, **if the universe were as old as Big Bang theorists contend, there should be little hydrogen in it. It would all have been transformed into helium by now. Yet stellar spectra reveal an abundance of hydrogen in the stars, therefore the universe *must* be youthful.**

Gravity

Gravity holds us firmly on the ground and keeps the Earth circling the Sun, and the Moon circling the Earth. This *invisible force* also draws down rain from the sky and causes the daily ocean tides.

It keeps the Earth in a spherical shape, and prevents our atmosphere from escaping into space. Cool stuff, hey?

Gravity is the weakest force in the universe, yet it is in perfect balance. If gravity were any stronger, the smaller stars could *not* form; any weaker, the bigger stars could *not* form and *no* heavy elements could exist. Only red dwarf stars would exist, and these would radiate too feebly to support life on a planet. **Think gravity happens by chance? Think AGAIN.**

Isaac Newton published his discoveries about gravity and motion in 1687, in his masterpiece *Principia.* He stated, **"Our most beautiful system of the Sun, planets, and comets could *only* proceed from the counsel and dominion of an *intelligent and powerful being.*"**
- *Creation.com*

Nuclear Force

It is the nuclear force that holds the atoms together. If it were larger, there would be *no* hydrogen, only helium and the heavy elements. If it were smaller, there would *only* be hydrogen and *no* heavy elements. **Without hydrogen and the heavy elements there could be *no* life.** Without hydrogen, there could be *no* stable stars.

If the nuclear force were only one part in a hundred stronger or weaker than it now is, carbon could *not* exist, and carbon is the basic element in *every* living thing. A two-percent increase would eliminate protons. Think *intelligent design. THINK GOD.* - *EvolutionFacts.com*

Earth's Magnetic Field

The **Earth's Magnetic Field** serves to deflect most of the solar wind, whose charged particles would otherwise strip away the ozone layer that   *protects* the Earth from harmful ultraviolet radiation.
- Wikipedia.org

The Earth's magnetic field is getting *weaker and weaker* with time. It *cannot* be millions of years old. The magnetic field, by the way, is one of the things that is not taken into consideration when performing *carbon dating;* because, the magnetic field is one of the things that prevents the formation of carbon 14 (C14). If the magnetic field is weakening, then we are getting more C14 than they had 5,000 years ago, which throws everything off.

As the Earth is spinning, it is gradually *slowing down* one thousandth of a second per day. The June issue of *Astronomy Magazine* announced, Be sure to add time to the clock. Earth's rotation is slowing down.

If the Earth is slowing down (and, it is), that of course means that it used to be going faster. If you go back in time 6,000 years, **Adam and Eve's day was a little *shorter* than ours.** If you want to claim that the Earth is millions or billions of years old, you have a problem.

You want to tell me that the dinosaurs lived 65 to 135 million years ago. If they did, I know what happened to them. They were blown off the surface of the Earth. The winds were 5,000 mph from the *Coriolis Effect*. The Earth was flattened out like a pancake. The days and nights would have been about 20 minutes each. No, you are mistaken. The Earth is not billions of years old as evolutionists claim. The slowing rate of the Earth proves that; it puts a time limit on it. Wake up, and smell the coffee. It *can't* be billions of years old. Now, you may need billions of years to make your theory look reasonable. The scientific facts say: No, it is *not* billions of years old.   - *Dr Kent Hovind (Creation Science Evangelism Seminar 1999)*

Impossible for Evolution

It would be *impossible* for Evolution to produce the delicate balances of these forces. They were planned. In spite of the delicate internal ratio balance within each of the four forces *(gravitation, electromagnetism, and the weak and strong forces),* those basic forces have strengths which differ so greatly from one another that the strongest is ten thousand, billion, billion, billion, billion times more powerful than the weakest of them. Yet the complicated math required for the Big Bang theory *requires* that *all* basic forces had to be the *same* in strength - during and just after that explosion occurred!

Evolutionists *cannot* claim that these delicate balances occurred as a result of *natural selection* or *mutations*, for we are here dealing with the basic properties of matter; there **is *no* room here for gradual "evolving."** The proton-neutron mass ratio, for example, is what it has *always* been - what it was since the beginning! It *has not* changed; it *will not* change. It began just right; there was *no* second chance! The same applies to *all* the other factors and balances in elemental matter and the physical principles governing them. - *EvolutionFacts.com*

"Once we see, however, that the *probability of life originating at random is so utterly minuscule* as to make it absurd, it becomes sensible to think that the favorable properties of physics on which life depends are in every respect deliberate... It is therefore almost inevitable that our own measure of intelligence must reflect higher intelligences, even to the limit of God, such a theory is so obvious that one wonders why it is not widely accepted as being self-evident. The reasons are *psychological* rather than scientific."
- *Sir Fred Hoyle (Evolution from Space)*

Proton to Neutron Ratio

A proton is a subatomic particle found in the nucleus of all atoms. It has a positive electric charge that is equal to the negative charge of the electron. A neutron is a subatomic particle that has *no* electric charge. The mass of the neutron must exceed that of the proton in order for the stable elements to exist. But **the neutron can *only* exceed the mass of the proton by an extremely small amount - an amount that is *exactly twice the mass* of the electron. That critical point of balance is only one part in a thousand.**

This did not evolve randomly... **If the ratio of the mass of the proton to neutron were to vary outside of that limit - *chaos would result.* If it were any less or more, atoms would fly apart or crush together - and *everything* would be *destroyed.*** If the mass of the proton were only slightly larger, the added weight would cause it to quickly become unstable and decay into a neutron, positron, and neutrino. This would *destroy hydrogen*, the dominant element in the universe. **A *Master Designer* planned that the proton's mass would be slightly smaller than that of the neutron. Otherwise the universe would collapse. *Evolution Mission Impossible.*** *- EvolutionFacts.com*

A *Neutron* walks into a bar and asks, "How much for a beer?" The bartender says, "For you? *No Charge.*"

When *Nothing* Makes Itself Into - *Something*

Go into a room and clean it out well. Remove all the furniture and even the dust. Seal up the windows and lock the doors and leave. Come back periodically and check to see what happens. **The *theory of evolution* expects the air inside the sealed room to *change itself into different types of matter over time;* such as birds, chemicals, grass, etc.** Or, take a vacuum bottle and extract as much air and gaseous material as possible. Seal it. Over time, the contents should change into something else, right? **Conclusion:** *NOTHING* NEVER makes itself into *Anything*. EVER. *- EvolutionFacts.com*

How Randomness Organizes Itself

Since this book is about THE *Fairytale for All Ages...* let's pretend a little (a *lot* actually)... Say you own a big antique store in *Make Believeland, USA.* Some thugs rob you one night and in doing so, smash virtually everything in the store into a million and one pieces.

You close shop and return once every year for the next 50 years to see if, let's just call it, *"Antique Evolution"* (which your twin brother Ricky told you is a real thing, and you never doubt Ricky, or his cousin Larry for that matter) is putting all those million and one pieces back together again, or better yet, putting them together to create newer and cooler things than before. Bigger and better things. Nothing? OK, let's let them sit there through many family generations, for millions of years. What? Still a million and one pieces?

Pretty absurd and random concept, right? Dream over. *Cue reality...*

Random Matter can *NEVER,* not in 50 years or even in a bazillion years... organize itself back into even something as simple as an antique vase or china set, let alone a *"living being," which of course is so much more complicated in every way.*

It's **completely impossible** for random, mindless *matter* to arrange itself into the universe. You know it, I know it, everybody knows it, yet many believe it, regardless. Our Sun, Moon, Earth and all living things on it could *only* come into existence by a *supernatural, intelligent designer,* and that is God, and Him alone.

Yes, *Evolution* is indeed a *Fairytale for All Ages.*

One Top Evolutionist Says That Evolution is "Anti-Knowledge"

"For over *20 years* I thought I was working on evolution... but there was not *one* thing I knew about it... So for the last few weeks I've tried putting a simple question to various people, the question is, 'Can you tell me *any* one thing that is true?' I tried that question on the Geology staff at the Field Museum of Natural History and the only answer I got was *silence.*

I tried it on the members of the Evolutionary Morphology Seminar at the University of Chicago, a very prestigious body of evolutionists, and all I got there was *silence* for a long time, eventually one person said, 'I do know one thing, it ought *not* to be taught in High School.'

Over the past few years.... you have experienced a *shift* from **evolution as knowledge to evolution as** *faith*… Evolution not only conveys *no knowledge,* but seems somehow to convey *anti-knowledge."*
- *Dr. Colin Patterson (Evolutionist, address at the American Museum of Natural History, New York City, Nov. 1981)*

Evolution's Latest Troubles

Recent scientific discoveries are forcing scientists to rethink the accepted version of evolution, known as the "Modern Synthesis." Evolutionists are looking for a *Third Way,* but these findings below fit nicely in the creation model.

1) Beneficial Mutations Are *Extremely* Rare: Beneficial mutations are rare to none and fail as a viable means to create complex new traits controlled by hundreds of interconnected genes. Without this magic mechanism, it is *impossible* to evolve fundamentally new types of creatures.

2) NO Big Pool of "Junk" DNA Exists: Virtually *all* of our DNA is now *proving to be functional.* Without vast regions of randomly evolving "junk" DNA, evolution *doesn't* have a place to work its magic.

3) Organisms Can Vary *Without* Genetic Change: Scientists are learning more and more about a mechanism called *epigenetics,* which does *not* require changes in the DNA sequence for creatures to vary in response to changes in the environment. These changes, which are heritable, were once thought to be related to genetic variability, so this finding further blunts a foundational premise of evolution.

4) Organisms Can Vary During Early Development: After an organism is conceived, its DNA is set, but development can be affected by something called *plasticity* and *environmental conditions.* This variability has been shown to be controlled by complex sensory-based recognition and response systems and *not* related to hypothetical evolutionary processes.

5) Organisms Depend on Other Organisms to Survive: We have always known that organisms live in highly complex interdependent communities with other organisms, but we're finding that this extends to interfacing with all sorts of microbes to survive. Humans, animals, and plants depend on many different types of microbes for proper growth, development, and daily living. This further complicates overly simplistic ideas of evolution.

- Dr. Jeffrey Tomkins at AnswersinGenesis.org

Death to Eugenics

Eugenics **is basically beliefs and practices for improving the human population by** *controlled breeding,* **by decreasing the occurrence of undesirable heritable characteristics, primarily in human beings. WHAT??**

Certainly not God's plan but many evolutionists have *no* problem with it. The thinking is, *if we all evolved from slime anyway, why not?*

It's essentially an *extension* **of Darwin's evolution hypothesis.** In fact, **Charlie's** *cousin Francis Galton* **is credited as the person who started it,** even though *Plato* suggested selective breeding over 2,000 years earlier. So, did Galton actually *plagiarize* it, and then only popularize it? Sound familiar? ***Looks like plagiarism evolved in the Darwin family.***

Some eugenicists also believe that abortion should be legal for the very purpose of promoting eugenics. **Abortions *murder* over 50 MILLION living, defenseless human babies (with a *heart beat*) around the world every *year*.** Many are *torn apart limb by limb* as they are pulled from their mother's womb. This is *barbaric.* **Try doing this to a baby elephant or gorilla in the womb and you'd end up in jail. What a sad, self-centered and demented world we live in where a mother's *rights* trump a defenseless human baby's *life*.**

Are the *mentally and physically handicapped*, and selected *senior citizens* of the world the next to be chosen for the death sentence?

LOVE

If you are of the evolutionary camp that thinks LOVE just *evolved* over time, then so be it. However, you're losing out on so much!!

Here's what the *Bible* says about LOVE... **"Let all that you do be done in love."** - *1 Corinthians 16:14* and **"We love because He** (God) **first loved us."** - *1 John 4:19*

And, from the "love chapter" heard at weddings... **"Love is patient, love is kind. It does not envy, it does not boast, it is not proud. It does not dishonor others, it is not self-seeking, it is not easily angered, it keeps no record of wrongs."** - *1 Corinthians 13:4-5*

When you get to know God, you really get to know His love for you and it's amazing!!! Something evolution cannot offer you.

TIME *(Evolution's Enemy)*

The evolutionist's mantra is, *"**Given enough time, anything can happen.**"* Oh really? First we must ask ourselves, is this a *true* statement? Does *time* really solve all evolutionary problems? Does it solve *any?* Is this just another *pickup* line for the theory? ;-)

In theory, it may sound good, but in reality... ***Long ages* are *not* evolution and they *cannot* produce evolution!** According to the theory, evolution can *only* occur by a sequence of production of *matter from nothing,* generation of *living organisms from non-living matter,* and evolution of living organisms into more advanced life forms by *natural selection or mutations.* And, even given billions and billions of years in which to do it, as we've seen, evolution *cannot* do any of that.

The evolutionist tells us that, *given enough time,* **all the** *insurmountable obstacles* **to spontaneous generation will somehow vanish and life can suddenly appear, grow, and flourish.** Oh really? If they only had some actual *proof* they might have a leg to stand on. Right now, *the evolution fairytale stands about as tall and strong as an old drunk pirate in a wheelchair. Arrr.*

Even split-second, continuous, multiple chemical activity (going on for ages) **and using all of the time and all of the space in the universe to carry on that activity could** *never* **accomplish what is needed.** It could *never* produce life out of absolutely *nothing.*

Just what is TIME, anyway? It's *not* **some magical substance. Time is merely a lot of past moments just like the present moment.** That's time in a nutshell.

To say that life originated in that seawater in some yesteryear, "because the sand and seawater was there *long enough,"* is just *wishful thinking* and nothing more.

THE <u>MORE</u> TIME, THE <u>LESS</u> LIKELIHOOD - Evolutionist G. Wald, in *"The Origin of Life,"* in the book, *Physics and Chemistry of Life,* asks *"Does time perform miracles?"* He then explains something that you and I will want to remember: **"If the probability of a certain event occurring is only 1/1000** (one chance in a thousand), **and we have sufficient time to repeat the attempts many times,** *the probability that it could happen would continue to remain only one in a thousand.* **This is because...** *PROBABILITIES have NO MEMORY!"* Friends, that is a powerful statement. So, what he's basically saying is that *probabilities always start from ground zero.*

THE TRUTH IS... **the** *longer the time,* **the greater the decay, and the** *less* **possibility that evolution could occur.**

WHERE ARE THE *REALLY OLD* **BUILDINGS?...** If human beings have been on this planet for over a million years, as theorized by evolutionists - then **we would have discovered a** *large* **amount of physical** *structures* **and** *written records* **extending back at least 100,000, if not 500,000 years or so. We do** *not.* **They only date back a** *few thousand* **years or so.** Evolutionists have *no* plausible explanation.

LONG AGES NEEDED... For nearly two centuries, evolutionists have known that, since there was *no proof* that evolution had occurred in the past and there was *no evidence* of it occurring today, they would need to postulate *long ages* as the means by which it somehow happened! - *EvolutionFacts.com (Arrr - pirate and other comments throughout the book were added by the author. ;-)*

"The belief that species are *immutable* (unchangeable) productions was almost unavoidable as long as the history of the world was thought to be of short duration." - *Charles Darwin (Origin of the Species... conclusion to his second edition)*

Get it? **Evolutionists are *FORCED to vehemently teach "long ages," or their little contrived fairytale *dies a fast death.***

The Food We Eat

Our bodies are designed to... *find, grow, harvest, eat, digest, and eliminate* **the food we eat.** Without food we would starve to death in weeks. Do you think all these super complex processes just somehow happened, and work together, *randomly?* Yeah right, then I've got a bridge to sell you. There's *no* way you can convince me that *bacon* is not from an all powerful and loving God. In fact, *bacon proves God exists.* ;-)

What Does *"Common Sense"* Say?

The definition of **Common Sense** is, a sound and prudent judgment based on a simple perception of the situation or facts.

Let me ask you a serious question... **do you really think humans *"evolved"* to do things like:** *talk, listen, see, smell, think, understand, cry, joke, create, sing, hope, write books and poetry, and feel emotions like joy, anger and love?*

Wouldn't *Intelligent Supernatural Creation* **make** *much* **more sense?** If you say, "no," then answer me, *why not?* Could evolution actually accomplish all of this, and so much more? Something to *think* *deeply* about, don't you think? Valid questions to ask yourself, and to research more.

Sure, these things should involve *mathematical probabilities* and real *scientific evidence,* but many times a little *common sense* and clear thinking goes a long way. **Evolutionists like to say that *"common sense"* proves their torpid theory, but it certainly does *not.***

Only *"Evidence"* can prove the theory, and that's where evolution *fails miserably.* Evolution is Dead, with *NO* chance of a resurrection. You might as well surrender to the facts of creation.

Communication Among Creatures

Evolutionists have *no explanation* for why *communication* occurs. ***Chance processes* couldn't have assembled the key ingredients needed for the elaborate messaging** we witness in the animal kingdom.

Higher animals routinely send forms of purposeful signals to influence the behaviors of other animals or even humans. To appreciate this, we must distinguish between animals using environmental cues, and those employing *communicative signals.*

A domesticated dog barks to alert humans, "I'm hungry! Feed me!" That barking is a communicative signal. Although simple, this is true communication. **There's a message sender, a receiver, and a transmitted message in the form of understandable coded information, and the sender's intention is to influence a responsive action by the receiver.**

When creature communication does occur - as it does worldwide, every day, in many contexts - it powerfully demonstrates God's providential bioengineering design for meaningful and purposeful messaging. *- ICR.org*

Evolution's Definition of... *STUPID*

"No one has ever found an organism that is known not to have parents, or a parent. *This is the strongest evidence on behalf of evolution.*" - Tom Bethell *(Agnostic Evolutionist, Harper's, February 1985, p. 81)*

WHAT?? This is the *strongest evidence* for evolution? Are you kidding me? Having *parents* makes evolution true? Many people will try to make some weird sense of it and will let completely nonsensical comments like this control their destiny and keep them from knowing the real creator of all parents and children. PARENTS do *not prove* evolution exists. Evolution is a JOKE, it's DEAD!! Stop following that *unenlightened* and *out-of-touch* crowd, if you are now.

And, **isn't it funny how skeptics say that *God is not real* because they can't *see* him - yet they say that *evolution is real,* which they have *not* seen or proved.** What sense does that make?

USE YOUR BRAIN... Stuff like this is crazy and makes *no* sense intellectually, whatsoever. None!! It's laughable. It's downright STUPID!! Don't accept foolish comments like that. Don't allow people to get away with idiotic comments like this as if they're true. These are an insult to your intelligence. *Challenge* them at every step when they make really dumb comments.

Start by asking them to explain their comments, and to provide *evidence,* if that's necessary.

"Fear not, for I am with you; be not dismayed, for I am your God; I will strengthen you, I will help you, I will uphold you with my righteous right hand."

- Isaiah 41:10

Animal Kingdom
(Following are Only a Few of Thousands of Examples That Disprove Evolution)

I can prove evolution wrong... *Humming Birds* have existed for thousands of years and they still haven't learned the *words* yet.
(Come on now, that's kind of funny. Laugh, it's good for you)

Survival of the Fittest?

The theory of evolution is based on the idea that, in any given environment, only certain organisms will succeed and all others will fail and die out.

SURVIVAL OF THE FITTEST... **has** *NOTHING* **to do with evolution... In fact, it actually accomplishes the exact** *OPPOSITE* **of evolution!** The hardships of life cull out the weakened forms of each species, and thus keep each species very stable. There is *nothing* in this process that has *anything* to do with evolution.

Species Evolution *NEVER* **occurs by means of** *Natural Selection.* Evolutionists have ransacked the plant and animal kingdoms for examples of cross-species evolution (by any means, natural selection or otherwise), and have been *unable* to find them. What they have found are some interesting examples of variations **WITHIN the SAME** species. These they present to the public and in school books as so-called "evidence" of evolution - which is *really bad science* (if you can even call it *science)*, and is COMPLETELY INCORRECT.

The *Monkey* is said to have developed a tail so it can climb trees better, but the *gibbon, manx cat and bear* climb trees and they *don't* have tails. The domestic cat climbs trees and has a tail, but does *not* use it for that purpose.

The *Horse* has uncrowned teeth, long legs, and a bushy tail so it will be *fit for survival (so says the theory).* The *cow* grazes in the same field and has crowned teeth, shorter legs, and a tail with a tuft on the end, and does just as well.

Why does the female *Duke of Burgundy Butterfly* walk on six legs, while its mate only walks on four?

Evolutionists say that *Plants* evolved berries to aid seed distribution by animals. Why then are some berries poisonous?

The *Queen Ant* produces worker ants which are sterile and thus unable to pass on improvements to offspring (*nor* receive them from their ancestors). How then could the worker bee evolve? The queen produces all the bees.

Cats **descend trees *tail first* -** but *leopards* survive just as well as the only member of the cat family that descends *head first.* Why then did the others *evolve* the pattern of going down tail first?

Evolutionists maintain that *Feathers* evolved for the purpose of flight. Why then do such birds as ostriches and penguins *not* fly? How can bats fly, when they have *no* feathers?

Why do *Insects* and *Birds* which are in identical environments have *different colors? - EvolutionFacts.com*

"Gregor Mendel's experiments clearly showed that one species could *not* transmute into another one. A genetic barrier existed

that could *not* be bridged. Mendel's work laid the basis for modern genetics; and his discoveries effectively *destroyed* the basis for species evolution." - *Michael Pitman (Evolutionist. Adam and Evolution, 1984)*

SOMETHING TO THINK ABOUT... if evolution says that *Survival of the Fittest* is true, **why don't most living creatures EAT THEIR BABIES? And, how would they know *not* to eat their young? You think if evolution were true the mom or dad would look at the fresh new food source as their next meal.** *"Hey hun, lunch is served. Come and get it."* **Sick thought but why in the world would they** *think or behave* **otherwise? Second thought, why then wouldn't the parents try to eat each other in the first place? Hey, as evolution likes to say, it's** *survival of the fittest.*

The Giraffe *(Evolution Crusher)*

"The *hypothesis* **that life has developed from** *inorganic matter* **is still an article of** *faith."* - *J.W.N. Sullivan (Evolutionist. The Limitations of Science (1933), p. 95)*

Charles Darwin wrote in his book, *On the Origin of Species,* that **the giraffe was just a** *regular animal* **that grew a long neck to reach the higher branches to eat and stay alive. Poor Charlie didn't have a clue,** and obviously didn't know much about giraffes! There is far more to a giraffe than merely "a long neck."

Most evolutionists still believe this utterly *ridiculous* theory even today. They are *forced* to, actually, or it disproves their theory and opens the door to a *Supernatural Creator.* They surely couldn't allow that to happen.

So, how then did other animals living among the giraffes; such as horses, elephants, sheep, hyenas, chickens, etc. - stay alive if they too could *not* reach high into the tree tops? If we use simple *logic* we see that of course the giraffe's neck *didn't* need to grow to keep them alive.

And, **they had to get their *water* from the *ground,*** but they couldn't get their food from the ground? Seriously? Hey, maybe there were *water trees.* HaHa. See why I said it's ridiculous? But wait, there's more...

The giraffe has the most powerful *heart* in the animal kingdom. This is due to the fact that it has *double* the normal blood pressure. This high blood pressure is required to pump blood all the way up to its brain.

The giraffe's blood pressure is two or three times that of a healthy man, and probably is the highest in the world. Because the giraffe has such a long neck (10-12 feet in length), its heart must exert an immense force to pump blood through the carotid artery to the brain. The giraffe's heart is huge; it weighs 25 pounds, is 2 feet long, and has walls up to 3 inches thick.

In contrast, the brain of any animal is a very delicate structure and is *not* able to stand high blood pressure. What happens when the giraffe bends over to take a drink from a pond? Obviously, we have here an *impossible* situation. High pressure is needed to get blood to the brain, yet that very pressure should destroy the brain when it lowers its head to the ground.

Four carefully thought-out *design* factors nicely solve this problem: **(1)** The giraffe has in his jugular veins a series *of one-way* check valves. These immediately close as soon as the head is lowered! But there is still a large amount of blood in the carotid artery; too much. **(2)** That extra blood is immediately shunted to a special spongy tissue, located near the brain and filled with small blood vessels, which *absorb* it. In addition, **(3)** the cerebrospinal fluid, which bathes the brain and spinal column itself, produces a *counter-pressure* to prevent rupture or capillary leakage. Last but not least, **(4)** the walls of the giraffe's arteries are thicker than those of any other mammal.

The giraffe is one of God's many incredibly *created* creatures. There's NO possible way for this all to *evolve* little by little over extended periods of time. *Think* about it. A giraffe evolving is completely impossible!

Evolution is a fairytale. Stop believing otherwise, if you still do. **The Theory of Evolution is not only *un*scientific and wrong, it's amazingly… STUPID!!** - *EvolutionFacts.com (Plus, some added content from the author of this book)*

So many of the so-called *proofs for evolution* from the *experts,* are so stupid they make debunking the theory as easy as *taking candy from a baby.*

The Beautiful Butterfly
(Evolution Crusher)

There are about 17,500 to 20,000 different *species* of butterflies in the world (725 to 750 in the U.S. alone). Somebody's got a lot of time on their hands to count all of these little beauties. Their beautiful wings are created by millions of minuscule scales that reflect specific wavelengths of light to produce their brilliant colors.

Butterflies go through an amazing transformation during their lifecycle. **The insect begins as an *egg,* then a crawling *caterpillar,* followed by the *pupa* stage** in which the caterpillar begins to convulse breaking off its outer skin, thus becoming a *chrysalis.* **Then, within the next day its organs disintegrate into a soupy *liquid.* Yes, LIQUID!! Miraculously after just a week or two, a *complex winged butterfly appears.* This metamorphosis takes place in a matter of *days*. Not millions of years. God is so incredible.**

The adult butterfly now has 6 legs, antennae, a specialized feeding tube, 2 amazing compound eyes, complex reproductive organs, and 4 ornate wings. The caterpillar had none of these features just *days* ago.

Clearly the *genetic instructions* for all these stages were programmed into the insect by the Creator from the very beginning of time.

"But ask the animals, and they will teach you, or the birds in the sky, and they will tell you; or speak to the Earth, and it will teach you, or let the fish in the sea inform you. Which of all these does not know that the hand of the Lord has done this? In his hand is the life of *every* creature and the breath of *all* mankind."
- Job 12:7-10

The Kangaroo *(Evolution Crusher)*

This is WILD... The kangaroo makes *"two kinds"* of milk *simultaneously*: milk for the *tiny baby,* AND other milk for a young kangaroo hopping alongside her. Each type of milk *differs considerably* in nutritional content. This definitely does *not* happen because of *evolutionary random chance processes.* NO WAY!!

After being born, the baby kangaroo journeys to its mother's pouch and begins nursing. After about 9 months it will begin climbing out of its mother's pouch and begin feeding. But, at times, it will jump back in and continue taking milk. Then, at about 10 months it no longer jumps in, but remains with its mother and reaches in from time to time to take more milk, until it is about 18 months old.

There are two striking facts about this: (1) The mother frequently has already given birth to *another* tiny baby which is also in the pouch nursing, so **she will have a baby *and* an adolescent nursing at the *same* time.** (2) The teat giving milk to the infant produces *different* milk than the one which the older one drinks from! **It matters not which teat it is; the older one will always receive a *different* composition of milk than the baby kangaroo is given.** The tiny infant has *very different nutritional needs.* But the question is how can the mother vary the type of milk which is given, at the *same* time, to both an adolescent and an infant kangaroo? How would it even know which one needs which formula? *Certainly not a random, wild guess.*

Bye Bye Evolution!! Hello Intelligent Design.

An example of this is the *red kangaroo,* which provides milk both to a tiny joey attached in the pouch to a teat, and also to a large joey which has left the pouch. The *older* one is given milk with a *33 percent higher proportion of protein* and a *400 percent higher proportion of fat.*

Even if evolution could do this, how in the world would it know that it had to be done this way? The evidence here for *intelligent design* is clearly seen, as the "theory" fades into the sunset.

BREAK TIME... OK, you've been presented a *lot* of information so far. If you are *still* an evolutionist... I'd respectfully like to ask you... WHY? What more do you need to *disprove* the theories of evolution?? Wherever you are right now in the process, I hope and pray that you will continue to read with an open mind and heart. (And, you believers, it's now time to crank up your prayers for these friends who are not quite on **God's Creation Team...** yet).

Desert Rat *(Evolution Crusher)*

Desert (Kangaroo) Rats in the Western U.S. can manufacture their own... WATER!! Oh, how we wish *we* could do it as inexpensively! Our worldwide water shortage is going to keep worsening. **This little rat does it be eating dry seeds, and then combining the *hydrogen* in them with *oxygen* from the air - and presto, nice, wet *water!*** They live 3 to 5 years without water. If they do drink water, they will die due to dehydration from all the nutrients they lose. - *EvolutionFacts.com*

A rat basically making water out of thin air. Something evolution could never accomplish. OK, maybe in a *fairytale.* I'll give you that. Another reason to believe in *Supernatural* Creation.

The Tiger Moth *(Evolution Crusher)*

The Tiger Moth has a *radar jamming device* which switches on as soon as a bat heads toward it, thus keeping the bat from locating it. Evolution at work? Yeah right.

The Trilobite *(Evolution Crusher)*

Trilobites are strange looking little creatures mostly found as fossils in the *lowest of the evolutionary strata,* called the *Cambrian.* However, **massive amounts of them have *also* been found *7,000 feet high in the mountains.*** Let that sink in. A *worldwide flood* is the *only* thing that could have placed them there in mass numbers. For the record, **a** lot of *other sea creatures* and *fish* have also been found *high in the mountains* around the world.

There are estimated to have been over 50 species of Trilobite. **All fossils found** (multiple billions) **in the Cambrian and actually** *every* **stratum above that, were found** *fully-formed.* This is called the *"Cambrian Explosion,"* where *every creature* found was *fully-formed and without any prior transitional/intermediate/in-between fossils ever found.* **Not just trilobites, but ALL of the fossils in the Cambrian.**

Trilobites had many very complex body parts. **Some did** *not* **even have eyes,** however, most had compound eyes, some with up to *15,000 lenses* **per eye.** I'd call that *fully-formed* from the beginning. **Virtually** *no* **fossils exist below this lowest (Cambrian) stratum.**

Trilobite Fossils (claimed to be *hundreds of millions* of years old by evolutionists) have been **found in HUMAN FOOTPRINTS,** in the Cambrian. The funny thing is, **they were found in the fossil of a human** *"sandal."* This has been confirmed by scientists. Evolutionists date the Cambrian layer to be around 500 *million* years old. The numbers don't add up for evolutionists, who claim humans are merely 200 *thousand* years old. Creationists say a few thousand years.

There are *numerous* **similar discoveries around the world, where** *human* **and** *multiple million year old animals and plants existed* *TOGETHER,* so this particular trilobite discovery mentioned is *not* some isolated incident or fraud. **It** *proves* **that trilobites (etc), which evolutionists claim to be among the oldest fossils to be discovered, actually lived with humans in recent modern days just** *thousands* **of years ago,** *not* **hundreds of millions of years ago as the silly evolutionary fable teller scientists like to shout from the rooftops.**

The Octopus *(Evolution Crusher)*

IN CAPTIVITY, the octopus is renowned for *unruly* **behavior - such as tampering with or blocking outlet valves, causing its tank to overflow.**

And, it can be very difficult to keep contained. **It can squeeze its** *boneless body* **through a space not much bigger than its eye.**

Institutions that have housed octopuses, tell of similar great escapes, and also of their overnight raids to catch and eat fish in other tanks. **One captive octopus learned to turn off electric lights by using its siphon/jet to squirt water at them, short-circuiting the power supply.** Another took an apparent dislike to a certain attendant, squirting a stream of salty water at her whenever she came within range, even remembering to squirt her after she returned from being absent for several months. Some see such behavior as **clear evidence of** *remarkable intelligence.*

Unquestionably, octopuses are *pretty good at sophisticated kinds of learning* such as being **able to open a screw-top jar from both the outside** *and* **inside.**

Their ability to **regenerate lost tissue, organs, and whole appendages** has attracted much research interest (the medical ramifications for *ourselves* could potentially be huge).

When an octopus dies, if not gobbled by scavengers, its soft body rapidly decays to be *little more than a slimy blob.*

Mark Purnell of the Palaeontological Association said that **finding a** *fossil* **octopus is about as unlikely as finding a fossil** *sneeze.*

Octopuses can *match* **their background (a** *coral reef* **for example) by** *instantly changing color* **so perfectly, even a keen observer loses**

sight of them. Watch a video of this online and you'll be blown away. Their skin has been described as being *"like a pixelated video screen,"* with the top layer containing tens of thousands of tiny pockets of different colors that can be independently opened and closed so as to exhibit the color scheme of the moment. Underlying that surface display is a layer of reflective cells stacked like a diffraction grating to create iridescence, with yet another layer beneath them to further bounce back incoming light. In the blue-ringed octopuses, the color is aposematic or warning, in this case that they have a highly venomous bite. No evolution in action here.

The camouflage is *not* static; the skin really can display flashing pulses of color like a *video screen*, seamlessly incorporating the passing shadows from clouds, the dappled shimmering of the seabed as waves scatter the sun's rays, or even forked lightning. **Added to all that, some octopuses can *shape-shift* so as to mimic the form and movement of other creatures** (the banded sea snake, and lion-fish) in an apparent effort to deter would-be predators. - *David Catchpoole (Creation Magazine - Vol. 41 No. 2). Full article at Creation.com*

The Mighty Octopus... Octopus fossils have been found in all different strata levels from the bottom up, and they all look identical to the modern day octopus. This makes sense with *Intelligent Creation*, but *not with evolution.* The octopus could never have evolved in a billion years. **They are indeed *Intelligent Design* at God's finest.**

The Happy Hippo *(Evolution Crusher)*

OK, maybe they aren't actually so happy but I wanted a fun title. In reality hippos (or *River Horse* in the Greek) can be extremely destructive. Their sheer size makes them virtually invulnerable to attack other animals (including crocodiles). In fact, **in Africa, more** *people* **are killed by hippos than by any other animal.** A large bull hippo may reach up to an incredible 4.5 m (15 ft) in length and weigh over 3.6 tonnes (8,000 pounds) - more than some mature female elephants!

The hippo's teeth are also fearsome. Its curving incisors are razor-sharp, and they can easily grow to over two feet in length. They are also composed of *ivory,* the same material as an elephant's tusks. But unlike the tusks of an elephant, hippo ivory doesn't turn yellow with age, making it a higher grade of ivory. In fact, **George Washington's teeth were actually made from *hippo ivory,* not wood as the tale goes.** - *Creation.com*

The San Diego Zoo has recorded **hippo grunts to hit about 115 decibels, comparable to a rock concert when you're about 15 feet away from the main speakers** (Where many *hippies* used to stand back in the day, to later in life go half-deaf).

And, last but certainly not least, hippos usually spend as much as 16 hours or so a day in the water. The reason being, their skin is very sensitive to sunlight, thus they secrete an oily reddish substance, which at one time was thought to be blood. This serves as an *antibiotic and a sunscreen* as well. While awake, they can hold their breath for up to five minutes. **While they *sleep* in the water, they *surface automatically and breathe without waking up.*** We have a very *Creative Creator,* wouldn't you agree?

If you choose to believe these things just *evolved by sheer accident* over long periods of time, then so be it. As if evolution would have a hippo know it needed to surface to get oxygen while it stayed *asleep.*

Come on now, really? Creation Science would paint a much *different* picture, such as this is how *God* created the Happy Hippo.

Amazing SeaHorse *(Evolution Crusher)*

Seahorses come in 45 species and usually live along the shore, among seaweed and other plants. They have only one mate, and generally don't travel more than a few meters. Their size varies from about 4 to 30 centimeters (1½–12 inches), and they continue to grow throughout their three years of life.

Study the seahorse carefully and you'll find strong and convincing *evidence* that points to it being wonderfully *designed* by God the Creator. Certainly *not* evolution.

Let's take a look at this amazing sea creature. A protective bony armor cleverly protects it from imminent danger. Their tough skeleton makes it unappetizing for predators, so seahorses are usually left alone.

The seahorse uses its fins to swim *vertically*, and rises or sinks by cleverly altering the volume of *gas* within its swim bladder. If this bladder is damaged, and it loses even a tiny bit of gas, it sinks to the bottom, where it will lie helpless until death.

If the seahorse is the product of evolution, *we must ask* how this creature managed to survive while its bladder evolved? The whole idea of the seahorse's complex bladder evolving by trial and error is unimaginable. Clearly, it is more reasonable to believe it was created through the work of the *Master Designer*.

Babies arrive by the *MALE*... Probably the most amazing, if not bizarre, aspect of the seahorse is that **the "male" gives birth to its live young.** This strange phenomenon has been known for only the past century or so.

The male has at the base of its abdomen, where it lacks armor plating, a large skin pouch and a slit-like opening. **The female lays the eggs**

into his pouch, where the male fertilizes them as they're deposited. She may continue laying eggs until his pouch is full, perhaps with as many as 600 eggs. The lining inside the pouch becomes sponge-like and filled with blood vessels which play some part in nourishing the eggs. This is an extraordinary characteristic of the male seahorse. Egg-laying complete, the dad-to-be swims off with his swollen pouch, *a living baby carriage.*

One or two months later he gives birth to tiny replicas of the adults. The little bundles of joy are squirted out until the pouch is empty.

Evolution is at a loss to account for the seahorse's reproductive functions. The whole process is simply too unorthodox. Indeed, the whole make-up of the seahorse is something of an enigma, if one tries to explain it as a product of evolution. As one authority said some years ago, "The *seahorse* is in a similar category with the platypus, as far as evolution is concerned: it presents an enigma that baffles and frustrates all theories that seek to account for it! Admit the *Divine Designer,* and all is accounted for."

Fossil Problem for Evolutionists... *Design* **is evident in the seahorse, but the** *"fossil record" is yet another problem* **for those who believe seahorses have evolved.** The evolutionist needs fossils showing a gradual development of lower animal life into the more complex seahorse to establish that the seahorse is the result of evolutionary processes over millions of years. Unfortunately for the evolutionist, *fossil seahorses are unknown.*

Like countless creatures of the sea, sky and land, *there is no link connecting the seahorse to any other form of life.* Like all other basic kinds of creatures, the complex seahorse appears to have been **created suddenly,** as the book of Genesis implies. - *Creation.com*

Mary River Turtle *(Evolution Crusher)*

The Mary River Turtle of Queensland, Australia has long **BRIGHT GREEN HAIR on its head**, which makes it look a bit like a *Punk Rocker* without his guitar. It's actually strands of algae which resemble hair.

Weird indeed, however, what's even more *special* is that it can… ***stay underwater* for 3 *DAYS*** by ***breathing through its... genitals.*** You can't make this stuff up. Yeah, God has a great sense of humor.

A Few More Evolution Crushers

Created *only* by an ***Intelligent Designer*** - How about a **COCKROACH** that has *GPS ability* in his brain similar to humans, or the **AFRICIAN PALM SWIFT bird** with a built-in *sensitive barometer* in its brain letting it know when storms are approaching, or the **GLASS FROG** which is literally *see-through* so you can see its organs through its skin, or the **PACU FISH** has *teeth that look identical to a human adult* (freaky looking), or the **AXOLOTL** which *regrows missing limbs,* and let's not forget about the *egg laying mammal* - the **PLATYPUS.** There are literally hundreds more examples of very weird creatures. Do you really think *random evolution* would, or could, be this creative and humorous? Nah. Me thinketh not. Let's give credit where credit is due, to God.

The animal kingdom is *extremely diverse* in so many ways. Do you really think they all were *not* created by an all powerful *intelligent designer*? Do you think that *unproven, unguided, random chance evolutionary processes could possibly create all this diversity?*

That's just a *wild fairytale* my friends and nothing more. Deep down we all (you included) know that evolution is a fable.

The Story of the Bear and the Atheist

So, this atheist is walking through the woods one day, enjoying all that **"Random Accident of Evolution"** has created. *"How wonderful,"* he says to himself.

While he's walking along the path, he hears some noise in the bushes. He looks around and sees a giant grizzly bear charging at him!

He runs up the path trying to get away, the bear now running close behind. His heart is pumping frantically as he tries to run faster. He looks over his shoulder as the bear closes in and grabs him in a big bear hug.

The atheist, cries out, *"Save me God, save me!"* God freezes time and says to the man...*"You denied my existence all your life and now you ask me to save you. Isn't that just a bit hypocritical?"*

The man thinks about it for a second and says, *"You have a good point there, God. OK, how about you spare my life this one time by making the bear a believer in you?"*

God says, *"OK, if that's what you want, then I'll do it for you."*

Next, the bear starts to pray... *"Dear Lord, thank you for this meal that you have so generously placed in my paws. Amen."*

QUOTES ABOUT GOD

Great Scientists and Philosophers, Even Some Nobel Prize Winners Who Believed in Our Creator God

SIR ISAAC NEWTON *(founder of Classical Physics and Infinitesimal Calculus)...* "From His true dominion it follows that the true God is a *living, intelligent and powerful Being;* and from His other perfections, that *He is supreme, or most perfect. He is eternal and infinite, omnipotent and omniscient;* that is, His duration reaches from eternity to eternity; His presence from infinity to infinity; He governs *all* things, and knows *all* things that are or can be done." (Newton 1687, *Principia*; see also Caputo 2000, 88).

GALILEO GALILEI *(founder of Experimental Physics)...* "When I reflect on so many profoundly marvelous things that persons have grasped, sought, and done, I recognize even more clearly that *human intelligence is a work of God,* and one of the *most excellent."* (Galileo, cited in Caputo 2000, 85)

NICOLAUS COPERNICUS *(founder of Heliocentric Cosmology)...* "To know the mighty works of God, to comprehend His wisdom and majesty and power, to appreciate, in degree, the wonderful working of His laws, surely all this must be a pleasing and acceptable mode of worship to the Most High, to whom ignorance cannot be more gratifying than knowledge." (Copernicus, as cited in Neff 1952, 191-192; and in Hubbard 1905, v)

JOHANNES KEPLER *(founder of Physical Astronomy and Modern Optics)*... "The World of Nature, the World of Man, the World of God, all three fit *together*. We see how God, like a human architect, approached the founding of the world according to *order and rule, and measured everything in such a manner."* (Kepler, cited in Tiner 1977, 172)

SIR FRANCIS BACON *(founder of the scientific inductive method)*... In the first chapter "Of Truth" of his *Essays* (1601), Lord Bacon wrote: "The first creature of God, in the works of the days, was the light of the sense; the last, was the light of reason; and his sabbath work ever since, is the illumination of his Spirit. First he breathed light, upon the face of the matter or chaos; then he breathed light, into the face of man; and still he breatheth and inspireth light, into the face of his chosen." (Bacon 1875)

RENE DESCARTES *(founder of Analytical Geometry and Modern Philosophy)*... In the beginning of his *Meditations* (1641) Descartes wrote: "I have always been of the opinion that the two questions respecting God and the Soul were the chief of those that ought to be determined by help of Philosophy rather than of Theology; for although to us, the faithful, it be sufficient to hold as matters of faith, that the human soul does not perish with the body, and that *God exists,* it yet assuredly seems impossible ever to persuade infidels of the reality of any religion, or almost even any moral virtue, unless, first of all, those two things be proved to them by natural reason. And since in this life there are frequently greater rewards held out to vice than to virtue, few would prefer the right to the useful, if they were restrained neither by the fear of God nor the expectation of another life." (Descartes 1901)

BLAISE PASCAL *(founder of Hydrostatics, Hydrodynamics, and the Theory of Probabilities)*... "Without *Jesus Christ* man must be in vice and misery; with *Jesus Christ* man is *free* from vice and misery; *in Him is all our virtue and all our happiness. Apart from Him there is but vice, misery, darkness, death, and despair."* (Pascal 1910, No. 545-46)

LORD KELVIN *(Thermodynamics and Energetics. Physics)...* "Do not be afraid to be free thinkers. If you think strongly enough, you will be forced by science to the belief in God." - *"The atheistic idea is so nonsensical that I cannot put it into words."* (Lord Kelvin, Vict. Inst., 124, p. 267, as cited in Bowden 1982, 218)

- Quotes from Nobelist.Tripod.com

JOIN US DAILY ON FACEBOOK
Please LIKE, COMMENT and SHARE
facebook.com/CharlieWasWrong

Evolutionists "Refuting" Evolution

"We stopped looking for monsters under our bed when we realized that they were inside us." *- Charles Darwin*

"Great is the power of steady misrepresentation." *- Charles Darwin*

"If it could be demonstrated that any complex organ existed, which could *not* possibly be formed by numerous, successive, slight modifications, *my theory would absolutely break down."*
- Charles Darwin (The Origin of Species)

"Why then is *not* every Geological formation and every stratum full of such *intermediate* links? Geology assuredly does *not* reveal any such finely graduated organic chain; and this, perhaps, is the *most obvious and serious objection* which can be urged *against* the theory."
- Charles Darwin (The Origin of Species)

"To suppose that the 'eye' could have been formed by natural selection, seems, I confess, absurd in the highest degree."
- *Charles Darwin*

"Often a cold shudder has run through me, and I have asked myself whether I may have not devoted myself to a *fantasy.*"
- *Charles Darwin (Life and Letters, 1887, Vol. 2, p. 229)*

And, this Nobel Prize winner and Atheist/Agnostic says...

"When it comes to the origin of life, we have *only two possibilities* as to how life arose. One is *spontaneous generation* arising to evolution; the other is a *supernatural creative act of God.* There is *no* third possibility... **Spontaneous generation was scientifically *disproved* one hundred years ago by *Louis Pasteur, Spellanzani, Reddy and others.* That leads us scientifically to only *one* possible conclusion — that *life arose as a supernatural creative act of God*... I will *not* accept that philosophically because I do *not* want to believe in God. Therefore, I choose to believe in that which I know is *scientifically impossible*, spontaneous generation arising to evolution.**" *(in the Scientific American, August, 1954).*
- *Dr. George Wald (Evolutionist, Professor Emeritus of Biology at the University at Harvard, Nobel Prize winner in Biology)*

"Most modern biologists, having reviewed with satisfaction the downfall of the spontaneous generation hypothesis, yet unwilling to accept the alternative belief in special creation, are left with *nothing.*"
- *Dr. George Wald (Evolutionist, Professor Emeritus of Biology at the University at Harvard, Nobel Prize winner in Biology)*

"Evolution is *unproved* and *unprovable*. We believe it, only because the only alternative is special creation, and that is unthinkable."
- *Sir Arthur Keith (Evolutionist and author of the foreword for the 100th edition of "Origin of the Species")*

"My attempts to demonstrate evolution by an experiment carried on for more than 40 years have completely failed. It is *not even possible* to make a *caricature* of an evolution out of paleobiological facts... The idea of an evolution rests on *pure belief."* **- Dr. Nils Heribert-Nilsson** *(Swedish botanist and geneticist, of Lund University)*

"Gregor Mendel's experiments clearly showed that one species could not transmute into another one. A genetic barrier existed that could *not* be bridged. **Mendel's work laid the basis for modern genetics; and his discoveries effectively *destroyed* the basis for species evolution."** **- Michael Pitman** *(Evolutionist. Adam and Evolution, '84, pp. 63-64)*

Here's a quote from 1861 by Adam Sedgwick about the reason why the lethargic theory of evolution was created.

"From first to last it is a dish of rank materialism cleverly cooked up ... And why is this done? For no other reason, I am sure, except to make us *independent* of a Creator." **- Adam Sedgwick** *(British geologist and Professor of Geology at Cambridge, England)*

"The only competing explanation for the order we all see in the biological world is the notion of *special creation."* **- Dr. Colin Patterson** *(Evolutionist and senior Paleontologist at the British Museum of Natural History, which houses 60 million fossils)*

"Darwin's evolutionary explanation of the origins of man has been transformed into a *modern myth,* to the detriment of scientific and social progress... The secular myths of evolution have had a damaging effect on scientific research, leading to distortion, to needless controversy, and to *gross misuse of science...* I mean the stories, the narratives about change over time. How the dinosaurs became extinct, how the mammals evolved, where man came from. These seem to me to be little more than *story-telling."* **- Dr. Colin Patterson**

(Evolutionist and senior Paleontologist at the British Museum of Natural History, which houses approximately 60 million fossils).

"The probability of life originating by *accident* is comparable to the probability of the unabridged dictionary resulting from an *explosion in a printing shop."* **- *Dr. Edwin Conklin*** *(Evolutionist and professor of biology at Princeton University)*

"I suppose the reason we leaped at the origin of species was because the idea of God interfered with our sexual mores." **- *Sir Julian Huxley*** *(President of the United Nations Educational, Scientific, Cultural Organization)*

"The German Fuhrer, as I have consistently maintained, is an *evolutionist;* he has consistently sought to make the practices of Germany *conform to the theory of evolution."* **- *Sir Arthur Keith*** *(Militant Anti-Christian physical anthropologist)*

"If I knew of any Evolutionary transitionals, fossil or living, I would certainly have included them in my book, *'Evolution.'"*
*- **Dr. Colin Patterson*** *(Evolutionist and senior Paleontologist at the British Museum of Natural History, which houses 60 million fossils)*

THIS ONE IS <u>POWERFUL</u> (and worth repeating again)...

"For over 20 years I thought I was working on evolution....But there was *not one thing I knew about it...* So for the last few weeks I've tried putting a simple question to various people, the question is, "Can you tell me any one thing that is true?" I tried that question on the Geology staff at the Field Museum of Natural History and the only answer I got was silence. I tried it on the members of the Evolutionary Morphology Seminar in the University of Chicago, a very prestigious body of Evolutionists, and all I got there was silence for a long time and eventually one person said, "Yes, I do know one thing, it ought *not* to be taught in High School"....over the past few years....you have experienced a shift from Evolution as knowledge to evolution as faith... **Evolution not only conveys** *no knowledge,* **but seems somehow to convey** *anti-knowledge."* **- Dr. Collin Patterson** *(Evolutionist, address at the American Museum of Natural History, New York City, Nov. 1981)*

"Humanism is the belief that man shapes his own destiny. It is a constructive philosophy, a non-theistic *religion,* a way of life."
- American Humanist Association

"If complex organisms ever did evolve from simpler ones, the process took place contrary to the laws of nature, and must have involved what may rightly be termed the *miraculous."* **- R.E.D. Clark** *(Victoria Institute, 1943, p. 63)*

"The more one studies paleontology, the more certain one becomes that **evolution is based on faith alone**... exactly the same sort of faith which it is necessary to have when one encounters the great mysteries of religion." **- Louis Trenchark More** *(Evolutionist quoted in Science and the Two-tailed Dinosaur, p. 33)*

"If I as a geologist were called upon to explain briefly our modern ideas of the origin of the Earth and the development of life on it to a simple, pastoral, people such as the tribes to whom the Book of Genesis was addressed, I could hardly do better than follow rather closely much of the language of the first chapter of *Genesis."* **- Wallace Pratt** *(quoted by W.L. Copithorne, in "The Worlds of Wallace Pratt," The Lamp, Fall 1971, p. 14)*

"Evolutionism is a *fairy tale* for grown-ups. This theory has helped *nothing* in the progress of science. *It is USELESS."* **- Bounoure, Le Monde et la Vie** *(Director of Research at the National Center of Scientific Research in France in 1983]*

"I am *not* satisfied that Darwin proved his point or that his influence in scientific and public thinking has been beneficial - the success of Darwinism was accomplished by a *decline in scientific integrity."*
- **W.R. Thompson** *(Evolutionist. Introduction to Charles Darwin, The Origin of the Species)*

"Unfortunately for Darwin's future reputation, his life was spent on the problem of evolution which is deductive by nature... It is absurd to expect that many facts will not always be irreconcilable with any theory of evolution; and, today, *every one of his theories is contradicted* by facts." - **T. Mora** *(Evolutionist. The Dogma of Evolution, p. 194)*

"I myself am convinced that the theory of evolution, especially the extent to which it's been applied, will be one of the *great jokes in the history books* of the future. Posterity will marvel that so *very flimsy and dubious an hypothesis* could be accepted with the incredible credulity that it has." - **Malcolm Muggeridge** *(Well-known British journalist and philosopher)*

"Creation and evolution, between them, exhaust the possible explanations for the origin of living things. *Organisms either appeared on the Earth fully developed or they did not* . . If they did appear in a fully developed state, they must have been created by some omnipotent intelligence." - **D.J. Futuyma** *(Science on Trial (1983), p. 197)*

"I think, however, that we must go further than this and admit that *the only acceptable explanation is creation.* I know that this is anathema to physicists, as indeed it is to me, but we must *not* reject a theory that we do not like if the experimental *evidence* supports it."
- **H. Lipson** *(Evolutionist. "A Physicist Looks at Evolution" Physics Bulletin 31 (1980), p. 138)*

"The reasonable view was to believe in spontaneous generation; the only alternative, to believe in a single, primary act of supernatural creation. There is *no* third position." **- George Wald** *(Evolutionist. "Origin of Life," Scientific American, August 1954, p. 48)*

How's this for ignorance (and, he's actually serious)...

"No one has ever found an organism that is known not to have parents, or a parent. This is the *strongest* evidence on behalf of evolution." **- Tom Bethell** *("Agnostic Evolutionists, "Harper's, February 1985, p. 81)*

And THIS... "If there were no *imperfections*, there would be no evidence to favor evolution by natural selection over creation." **- Jeremy Charles** *(Evolutionist. "The Difficulties of Darwinism," New Scientist, Vol. 102 (May 17, 1984), p. 29)*

"The theory of evolution (is) a theory universally accepted *not* because it can be proved by logically coherent evidence to be true but because the only alternative, special creation, is clearly unimaginable." **- D.M.S. Watson** *(Evolutionist. "Adaptation," Nature, Vol. 123 p. 233 - 1929)*

"Today our duty is to destroy the myth of evolution, considered as a simple, understood and explained phenomenon which keeps rapidly unfolding before us. Biologists must be encouraged to think about the weaknesses and extrapolations that the theoreticians put forward or lay down as established truths. The deceit is sometimes unconscious, but not always, since some people, owing to their sectarianism, *purposely overlook reality and refuse to acknowledge the inadequacies and falsity of their beliefs."* **- Pierre-Paul Grease** *(Evolutionist. Evolution of Living Organisms (1977), p. 8)*

"'Scientists who go about teaching that evolution is a fact of life are great con men, and the story they are telling may be the *greatest hoax*

ever. In explaining evolution - *we do not have one iota of fact.'* [Tahmisian called it] a tangled mishmash of guessing games and figure juggling." **-** ***Fresno Bee*** *(August 20, 1959, p. 1-B, quoting evolutionist T.N. Tahmisian, physiologist for the Atomic Energy Commission)*

"Scientists have *no* proof that life was *not* the result of an act of *creation.*" **-** ***Robert Jastrow*** *(Evolutionist. The Enchanted Loom: Mind in the Universe (1981), p. 19)*

"To my mind, the theory does *not* stand up at all." **-** ***H. Lipson*** *(Evolutionist. "A Physicist Looks at Evolution," Physics Bulletin 31 (1980), p. 138)*

"The *hypothesis* that life has developed from *inorganic matter* is still an article of *faith." -* ***J. W.N. Sullivan*** *(Evolutionist. The Limitations of Science (1933), p. 95)*

"In accepting evolution as fact, how many biologists pause to reflect that science is built upon theories that have been proved by experiment to be correct, or remember that the theory of animal evolution has *never* been thus proved." **-** ***L.H. Matthews*** *(Evolutionist. "Introduction," Origin of Species, Charles Darwin (1971 edition)*

"Unfortunately, in the field of evolution most explanations are *not* good. As a matter of fact, they hardly qualify as explanations at all; they are *suggestions, hunches, pipe dreams, hardly worthy of being called hypotheses." -* ***Norman Macbeth*** *(Darwin Retried - 1971)*

"Ultimately, the Darwinian theory of evolution is no more nor less than the great cosmogenic *myth* of the twentieth century." **-** ***Michael Denton*** *(Evolutionist. Evolution: A Theory in Crisis (1985), p. 358)*

"The origin of all diversity among living beings remains a mystery as totally unexplained as if the book of Mr. Darwin had never been written, for *no theory unsupported by fact,* however plausible it may appear, can be admitted in science." **- L Agassiz** *(on the Origin of Species, American Journal of Science 30 (1880), p. 154)*

"Such explanations tend to fall into one or the other of two broad categories: special creation *or* evolution. Various admixtures and modifications of these two concepts exist, but it seems impossible to imagine an explanation of origins that lies completely outside the two ideas." **- Davis and E. Solomon** *(Evolutionist. The World of Biology (1974), p. 395)*

Darwin's biggest disciple and bulldog, Thomas Huxley, admitted that "creationism is in no way illogical." When an explanation is logical and has the evidence in its favor, and the rival theory is illogical and lacks any substantive evidence, then the conclusion is rather obvious.

"'Creation' in the ordinary sense of the word, is perfectly conceivable. I find *no* difficulty in conceiving that, at some former period, this universe was not in existence; and that it made its appearance in six days... in consequence of the volition of some pre-existing Being." **- Leonard Huxley** *(Evolutionist. Life and Letters of Thomas Henry Huxley, Vol. 11 (1903), p. 429)*

"Within the period of human history we do not know of a single instance of the transformation of a single species into another one." **- Dr. T.H. Morgan** *(Nobel Prize winner / Evolutionist)*

"Evolution is Religion." **- Michael Ruse** *(Evolutionist and past professor of philosophy and zoology at Guelph University)*

"Once we see, however, that the probability of life originating at random is so *utterly minuscule* as to make it *absurd,* it becomes sensible to think that the favorable properties of physics on which life depends are in every respect deliberate... It is therefore almost inevitable that our own measure of intelligence must reflect - higher intelligences - even to the limit of God - such a theory is so obvious that one wonders why it is not widely accepted as being self-evident. The reasons are *psychological* rather than scientific."
- Sir *Fred Hoyle* *(from "Evolution from Space")*

"The belief that life on Earth arose spontaneously from non-living matter, is simply a matter of *faith* in strict reductionism and is *based entirely on ideology."* - **Hubert P. Yockey** *(Non-Creationist. Information Theory and Molecular Biology, Cambridge University Press, UK, p. 284. 1992)*

"I have seen *NO evidence* whatsoever that these [evolutionary] changes can occur through the accumulation of gradual mutations."
- *Lynn Margulis* *(Famous evolutionary biologist. Science Vol. 252, 19 April 1991, p. 379. Note: She has been awarded membership in the National Academy of Sciences, and is the **ex-wife of Carl Sagan**)*

"Some *super-calculating intellect must have designed the properties of the carbon atom,* otherwise the chance of my finding such an atom through the blind forces of nature would be utterly minuscule."
- *Sir Fred Hoyle* *(Famous Astronomer who formulated the theory of stellar nucleosynthesis)*

"Evolution is a *fairy tale* for adults." **- *Jean Rostand*** *(Agnostic, and famous French biologist)*

"*Life could not arise spontaneously in a primeval soup of any kind...* Furthermore, ***no* geological evidence indicates an organic soup ever existed on this planet. We may therefore with fairness call this**

scenario *the Myth of the Pre-Biotic Soup.*" - *Sir Fred Hoyle* *(Famous Astronomer who formulated the theory of stellar nucleosynthesis. Non-Creationist)*

"It is *no more likely* that our world has evolved out of chaos than that a hurricane, blowing through a junkyard, should create a Boeing 747."
- Sir Fred Hoyle (Famous Astronomer who formulated the theory of stellar nucleosynthesis)

"Despite the bright promise that paleontology provides a means of 'seeing' evolution, it has presented some nasty difficulties for evolutionists, the most notorious of which is the presence of 'gaps' in the fossil record. Evolution requires *intermediate* forms between species and paleontology does *not* provide them..." *- David B. Kitts (Ph.D. -- Zoology, Head Curator, Department of Geology, Stoval Museum, and well-known evolutionary paleontologist. Evolution, Vol. 28, Sept. 1974)*

DEAR ATHEIST/AGNOSTIC/SKEPTIC... Are these atheists (evolutionists) the people you really want to build your trust and life around? God has so much more for you. Yes, YOU.

Creationists

"He who thinks half-heartedly will *not* believe in God; but he who really thinks *has* to believe in God." *- Sir Isaac Newton*

"The Universe, made for us by a supremely good and orderly *Creator.*" *- Nicolaus Copernicus*

"Mathematics is the language in which *God* has written the universe."
- Galileo

"Our prayers should be for blessings in general, for *God* knows best what is good for us." – *Socrates*

"God is *truth* and light his shadow." - *Plato*

"I love to think of nature as an unlimited broadcasting station, through which *God speaks to us every hour,* if we will only tune in."
- *George Washington Carver*

"The more I study nature, the more I stand amazed at the work of the *Creator.*" - *Louis Pasteur*

"By God, I understand, a substance which is infinite, independent, supremely intelligent, supremely powerful, and which created *everything* that exists. All these attributes are such that, the more carefully I concentrate on them, the less possible it seems that they could have originated from me alone. So, from what has been said *it must be concluded that God necessarily exists.* " - *Rene Descartes*

"*God has given you one face,* and you make yourself another."
- *William Shakespeare*

"Never lose an opportunity of seeing anything beautiful, for *beauty is God's handwriting.*" - *Ralph Waldo Emerson*

"The *evolution theory* is purely the product of the *imagination.* "
- *Sir Dr. Ambrose Flemming* (*Father of modern electronics*)

"The nearer I approach the end of my pilgrimage, the clearer is the *evidence* of the divine origin of the Bible, the grandeur and sublimity of God's remedy for fallen man are more appreciated, and the future is illumined with *hope and joy.* " - *Samuel Morse* (*famous for the invention of the telegraph*)

"That there is a *Supreme Intelligence I do not doubt.*"
- *Thomas Edison*

"The mathematical precision of the universe reveals the mathematical mind of *God."* - *Albert Einstein*

"I believe in Christianity as I believe that the sun has risen: not only because I see it, but because *by it I see everything else." - C.S. Lewis*

"Evolution is *not* a fact. It *doesn't even qualify as a theory or hypothesis.* It is a metaphysical research program, and is *not* really testable science." - *Karl Popper* *(science philosopher)*

"God heals, and the Doctor takes the Fees." - *Ben Franklin*

"Hypothesis [evolution] based on *no* evidence and irreconcilable with the facts.... These classical evolutionary theories are a gross over-simplification of an immensely complex and intricate mass of facts, and *it amazes me that they are swallowed so uncritically and readily, and for such a long time, by so many scientists without a murmur of protest."* - *Sir Ernst Chan* *(Nobel Prize for penicillin)*

"The pathetic thing about it is that many scientists are trying to prove the doctrine of evolution, which *no* science can do." - *Dr. Robert A. Milikan (Physicist and Nobel Prize winner)*

"Those supposedly omniscient scientists who still teach evolution as though it were fact are finally seen for what they are... *frail men willing to believe a lie because it helps them avoid the truth."* Adding, "Faith is the substance of fossils hoped for, the evidence of links unseen." by Dr. Scott Huse (Collapse of Evolution) - *A. Lund*

"Surely, God on high has not refused to give us enough wisdom to find ways to bring us an improvement in relations between the two great nations on Earth." *- Mikhail Gorbachev*

"I learned to put my *trust in God* and to see Him as my strength."
- Rosa Parks

"The Good News borne by our risen Messiah who chose not one race, who chose not one country, who chose not one language, who chose not one tribe, *who chose all of humankind!" - Nelson Mandela*

"We are not some casual and meaningless product of evolution. *Each of us is the result of a thought of God."* *- Pope Benedict XVI*

"To say that a scientist can disprove the existence of God is like saying a mechanic can disprove the existence of Henry Ford. It doesn't follow." *- Frank Turek*

"The delusive theory of evolution has *no evidence* and is a flat-out sham, one of the *biggest frauds* ever pulled off on mankind. It must be stopped." *- Nick Hetcher*

"Evolution may fool some, but if you take a closer look you'll discover it's nothing more than a F*airytale for All Ages,* with *no* evidence to back up its myriad of lies. Come to think of it, it's more like a Never Ending *Nightmare." - Nick Hetcher*

"Evolution, Evilution. Potato, Pototo." *- Nick Hetcher*

"EVOLUTION IS DEAD!! Charlie Was WRONG. Prove me wrong. Show me the *transitional* fossils. Oh yeah, there aren't any."
- Nick Hetcher

"If evolution says that *Survival of the Fittest* is true, why don't most living creatures EAT THEIR BABIES? Or, their mates for that matter? I guess they would only have their friends over for dinner, ONCE! And, how would they know *not* to eat their young? If evolution were true, mom and dad would look at the fresh new food source as their next meal. *"Hey hun, lunch is served. Come and get it."* Sick thought but if evolution is true, why in the world would they *think or behave* otherwise? *- Nick Hetcher*

"Arrr… The evolution fairytale stands about as tall and strong as an old drunk pirate in a wheelchair." *- Nick Hetcher*

"Evolution is *Science Fiction* at it finest." *- Nick Hetcher*

"*Monkey Science* is for for monkeys, not humans." *- Nick Hetcher*

"Evolution offers *no* hope. God promises hope in abundance. Choose God! Choose Jesus." *- Nick Hetcher*

"In regards to creation; by disproving the *natural,* we prove the *supernatural.* There are no other options." *- Nick Hetcher*

Business People

"I believe God is managing affairs and that He doesn't need any advice from me. With *God* in charge, I believe everything will work out for the best in the end." *- Henry Ford*

"God gave me my money. I believe it is a religious duty to get all the money you can, fairly and honestly; to keep all you can, and to give away all you can." *- John D Rockerfeller*

"If a man's business requires so much of his time that he cannot attend the services of his church, then that man has more business than God intended him to have." **- JC Penney**

"After women, flowers are the most lovely thing God has given the world." **- Christian Dior**

"To say that it (the world) was generated by *random numbers,* that does seem, you know, sort of an uncharitable view. I think it makes sense to believe in God. **- Steve Jobs**

Sam Walton was a Christian but I couldn't find any *quotes* he said about God. I did find that as a man of faith, he taught Sunday school at his local Presbyterian church, and Christian values to Walmart employees.

"I have offended God and mankind because my work didn't reach the quality it should have." **- Leonardo da Vinci**

"I believe in God, family, and McDonald's." **- Ray Kroc**

"I pray to the Almighty to encourage you to dedicate your life to *Jesus Christ.* Whatever difficulties you may have to face, turn to him, recognize him and render him all your glory, and he will help you overcome." **- Colonel Sanders**

Celebrities

"I'm not the king. *Jesus Christ is the King.* I'm just an entertainer."
- Elvis

"Real men need Jesus. If your soul needs healing, the prescription you need is *not* Chuck Norris' tears, it's Jesus' blood." **- Chuck Norris**

" Put *God* first in everything you do." **- Denzel Washington**

"*God is real. God loves you,* God wants the best for you. Believe that, I do." **- Chris Pratt**

"I'm a Christian. Before I go onstage every night, I pause and dedicate the performance to *God,* in the sense of, 'Allow me to surrender.'"
- Hugh Jackman

"I believe that *Jesus* was the Son of God." **- Bono *(U2)***

"I am a Christian, and *I love the Lord."* **- Carrie Underwood**

"I pray to be a good servant to *God."* **- Mark Wahlberg**

"I love comedy. *God* has given me this platform." **- Jeff Foxworthy**

"I've always believed in *God and Jesus.*" **- Kevin Sorbo** *(TV's Hercules and star of the movie God's Not Dead)*

"*God* gave you a brain. Do the best you can with it."
- Clint Eastwood

"Do I believe in *God? Yes, I do.* I do." **- Anthony Hopkins**

"God gave me *life* - to continue to do things that I would never have done." *- Stevie Wonder*

"I don't stand for black man's side, I don't stand for white man's side, I stand for *God's* side." *- Bob Marley*

"It is rare that you get the opportunity to make a film about something that has a great Christian message." *- Reese Witherspoon*

"Jesus is the inspiration for anyone to go the distance."
- Sylvester Stallone

Top Athletes

"When I found *Jesus Christ,* I learned to be a better athlete. I wanted to prove to other boxers that you can take off this killer instinct stuff, you can be a great athlete, a great boxer, and love your brother." *- George Foreman (Oldest boxer to win the world boxing heavyweight championship. And, a well known celebrity)*

"My relationship with *Jesus Christ* is the most important thing in my life." *- Tim Tebow* *(former NFL football player)*

"People consider me a success because I'm a good football player and make lots of money. But if my heart's not right, if I'm not living a life pleasing to *God,* I'm a failure." *- Reggie White* *(NFL legend)*

Note from author: My family attended the same church in Green Bay as Reggie. He became my "MVC" (Most Valuable CheeseHead) in 1995 with a small t-shirt company I owned called Cheese Wear. He had a great sense of humor and was wonderful man of God.

"There's more to me than just this jersey I wear, and that's *Christ living inside of me.*" - **Stephen Curry** *(NBA super star)*

"I came from nothing into something and *I owe everything to God.* He gave me this blessing. It's all credit to the Lord." - **Manny Pacquiao** *(the only boxer to win twelve major world titles in eight different weight divisions)*

"I can sum it up like this: Thank *God* for the game of golf."
- **Arnold Palmer**

"Talent is *God* given. Be humble. Fame is man-given. Be grateful. Conceit is self-given." - **John Wooden** *(Superstar basketball coach)*
"God created the Heavens, he created the Earth! He created all the Hulkamaniacs! Then, he created this set of 24-inch pythons!"
- **Hulk Hogan** *(LOL, talking about his huge biceps)*

US Presidents

"It is impossible to rightly govern a nation without *God and the Bible.*" - **George Washington**

"We recognize no sovereign but *God,* and no King but *Jesus!"*
- **John Adams**

"The reason that Christianity is the best friend of government is because *Christianity is the only religion that changes the heart."*
- **Thomas Jefferson**

"I believe *the Bible is the best gift God has given to man.* All the good from the Savior of the world is communicated to us through this book." - **Abraham Lincoln**

"God gave us Abe Lincoln and Liberty, let us fight for both."
- Ulysses S Grant

"A thorough knowledge of the *Bible* is worth more than a college education**." *- Theodore Roosevelt***

"The rights of man come not from the generosity of the state but from the hand of *God.*" *- John F Kennedy*

"Within the covers of the *Bible* are all the answers for all the problems men face. - Live simply. Love generously. Care deeply. Speak kindly. Leave the rest to *God.*" *- Ronald Reagan*

"May He guide us now. And may *God* continue to bless the United States of America." *- George W Bush*

"*God is in control*—and my main responsibility is to love God with all of my heart, soul, and mind." *- Barack Obama*

"May we as Americans never forget the power of prayer and the greatness of our *Creator.*" *- Donald Trump*

Other Issues with Evolution

The March of Progress

The March of Progress, properly called *The Road to Homo Sapiens*, is a textbook *illustration* that presents 25 *million* years of human evolution. You know, it's that chart with a line of monkeys to apes to humans. *No evidence whatsoever,* yet sold to children and teenagers in school textbooks as if it actually happened. The day this would ever happen is the day monkey's fly. Ain't gonna happen.

If evolution is true, we would absolutely still see many half (whatever) / half (whatever) species living among us today!! *Transitional* **species. However, we** *only* **see** *fully-formed* **living creatures, just as we** *always* **have. No evolution has occurred.**

The Bogus Embryo Drawings

Then there's Darwin's most zealous disciple in the late 1800s, *Ernst Haeckel,* who first championed the idea of vestiges. He's the guy who constructed a chart (*simple hand sketches*) with human and animal embryos all looking *identical* with *gills and all,* which was later discovered to be *completely falsified.* He *faked* them to support an alleged law he made up called, the *Law of Recapitulation or Biogenetic Law.* It was *proven to be a complete sham* yet passed off as real at the time, and even made it into modern day textbooks.

By the mid-20th century, reputable scientists recognized that Haeckel's theory was *ridiculous* **and** *without any scientific basis.* **He**

was later **repeatedly** *charged and convicted* of *fraud* by several professors in a university court.

Yet, even though proven a complete fraud-job in the late 1800s, these *bogus pencil drawings* **have still been found in modern school textbooks almost 150 years later, as recent as 2013.** **Just one more** *verifiable proof* **that evolution is** *nothing more than a monstrous scam on society.* **Come on people, brighten up. Don't these kinds of things just insult your intelligence? They sure should.**

EMBRYONIC SIMILARITIES... The concept of *recapitulation* is based on the fact that there are similarities among embryos of people, animals, reptiles, birds, and fish.

It is true that embryonic similarities do indeed exist. Babies, before they are born, look quite a bit alike during the first few *weeks.* This includes human babies, raccoon babies, robin babies, lizard babies, and goldfish babies. They all *begin* as *extremely tiny round balls,* so of course, on the *outside* they look somewhat similar.

RECAPITULATION - Reading in scientific books, you will come across the word, *"recapitulation,"* **the theory that human embryos are really little better than the left-over parts of fish, chickens, lizards, and other animals.**

Did you ever notice that *big words* **are sometimes used as proof in themselves? "Because it is a big word, therefore it must be true."** But these big words only cover over a *very foolish and contrived theory.*

TOTALLY UNIQUE - Each of us began as something as small as a dot on a word on this page. Yet if we examine that almost microscopic egg, we find that human dot has totally different genes and chromosomes than the egg of any other type of animal or plant. Only the *outside* **appearance may be somewhat similar to**

that of other embryos. As it grows, its structures will continue to become more and more diverse from those of any other kind of plant or animal. *Every* species of animal *and* plant in the world has blood cells different from *all* others, and a **totally *unique* DNA code.**

The recapitulation theory is just too shallow *(and not true)* to really explain much of anything. Only *Creation* can explain what we see about us in nature. **The similarities found in embryos point to a single Creator, *not* to a common ancestor.** - *EvolutionFacts.com*

FUNNY THING is that both of these *so-called evidence for evolution* topics *(The March of Progress and The Bogus Embryo Drawings)* are *not* actually real, but rather *PENCIL SKETCH DRAWINGS!!* Pretty sketchy "theory," yet unsuspecting people literally buy into these cheesy sales pitches for evolution. But *you're* smarter than that, right? RIGHT? I just knew you were.

The Horse Series

30 DIFFERENT HORSES... In the 1870s, evolutionist Othniel C. Marsh claimed to have found **30 different kinds of horse fossils in Wyoming and Nebraska. He reconstructed and arranged these fossils in an evolutionary series, and they were put on display at Yale University. Copies of this "horse series" are to be found in many museums** in the United States and overseas. Visually, it *looks* convincing. That's the extent of the reality of it.

Funny how *World Book Encyclopedia* makes this next statement without *any actual evidence*, whatsoever.

"Horses are among the best-documented examples of evolutionary development." - World Book Encyclopedia (1982 ed.), p. 333.

FOURTEEN FLAWS IN THE SERIES... When we investigate this *so-called* **"horse series" carefully, we come upon 14 distinct** *problems* that negate the possibility that we have here a genuine series of evolved horses. We discover that **the evolutionists have merely selected a variety of different size animals, arranged them from small to large, and then called it all "a horse series."** This is pretty typical of most evolution propaganda.

1 - *Different animals in each series.* In the horse-series exhibit we see a small, three-toed animal that grows larger and becomes our single-toed horse. But **the sequence varies from museum to museum** (according to which non-horse smaller creatures have been selected to portray "early horses"). **There are over 20 different fossil horse series exhibits in the museums—with no two exactly alike!** The experts select from bones of smaller animals and place them to the left of bones of modern horses, and, presto! another horse series!

2 - *Imaginary, not real.* The sequence from small many-toed forms to large one-toed forms is completely absent in the actual fossil record. **Some smaller creatures have one or two toes; some larger ones have two or three.**

3 - *Number of rib bones*. **The number of rib bones does** *not* **agree with the sequence.** The four toed *Hyracothedum* has 18 pairs of ribs; the next creature has 19; there is a jump to 15; and finally back to 18 for *Equus,* the modern horse.

4 - *No transitional teeth.* **The teeth of the "horse" animals are either grazing or browsing types. There are no transitional types** of teeth between these two basic types. Of course there should be.

5 - *Not from "in-order" strata.* **The "horse" creatures do not come from the "proper" lower-to-upper rock strata sequence.** (Sometimes the smallest "horse" is found in the highest strata.)

6 - *Calling a badger a horse*. The first of the horses has been called *"Eohippus"* (dawn horse), but experts frequently prefer to call it *Hyracotherium,* since it **is like our modern *hyrax,* or *rock badger.* Some museums exclude Eohippus entirely because it is *identical* to the rabbit-like hyrax (daman) now living in Africa.** (Those experts who cling to their "Eohippus" theory have to admit that it climbed trees!) The four-toed Hyracotherium does not look the least bit like a horse. (The hyrax foot looks like a hoof, because it is a ***suction cup*** so the little animal can walk right up vertical trees! For the record, horses do *not* have suction cups on their feet!)

"The first animal in the series, Hyracotherium *(Eohippus), is so* different from the modern horse and so different from the next one in the series that there is a big question concerning its right to a place in the series . . [It has] a slender face with the eyes midway along the side, the presence of canine teeth, and not much of a diastema (space between front teeth and back teeth), arched back and long tail." - *H.G. Coffin (Creation: Accident or Design? (1969), pp. 194-195)*

7 - *Horse series exists only in museums*. A complete series of horse *fossils* in the correct evolutionary order has *never* been found anywhere in the world. **The fossil-bone horse series starts in North America (or Africa; there is dispute about this), jumps to Europe, and then back again to North America.** When they are found on the same continent (as at the John Day formation in Oregon), **the three-toed and one-toed are found in the same geological horizon (stratum).** Yet, according to evolutionary theory, it required millions of years for one species to make the change to another. Busted again.

8 - *Each one distinct from others*. There are *no* transitional forms between each of these "horses." As with all the other fossils, each suddenly appears *fully-formed* in the *fossil* record. This is just another good reason we need to trash the deceiving theory of evolution.

9 - *Bottom found at the top*. Fossils of Eohippus have been found in the top-most stratum, alongside of fossils of two modern horses: *Equus nevadensls* and *Equus accidentalis*.

10 - *Gaps below as well as above*. Eohippus, the earliest of these "horses," is completely unconnected by any supposed link to its presumed ancestors, the *condylarths*.

11 - *Recent ones below earlier ones*. In South America, the one-toed ("more recent") is even found below the three-toed ("more ancient") creature.

12 - *Never found in consecutive* stratum. Nowhere in the world are the fossils of the horse series found in successive stratum.

13 - *Heavily keyed to size*. The series shown in museum display generally depict an increase in size; and yet **the range in size of living horses today, from the tiny American miniature ponies to the enormous shires of England, is as great as that found in the fossil record.** However, the modern ones are all solidly *horses*.

14 - *Bones, an inadequate basis*. In reality, one cannot go by skeletal remains. Living horses and donkeys are obviously different species, but a collection of their bones would place them all together.

Fear to Speak Out Against Evolution

FEAR TO SPEAK... **Even though scientists may personally doubt the evolutionary theories and the *so-called evidence* for them, publicly most fear to tell the facts, lest it recoil on their own *salaried positions*.** One fossil expert, when cornered publicly, hedged by saying the horse series "was the best available example of a supposed transitional sequence." **It is a devastating *so-called* fact that the best available example is a - *carefully fabricated FAKE*.**

Dr. Eldredge [curator of the Department of Invertebrates of the American Museum of Natural History in New York City] called the textbook characterization of the horse series 'lamentable.' Thus, *deplorably bad.*

When scientists speak in their offices or behind closed doors, they frequently make candid statements that sharply *conflict* with statements they make for public consumption before the media. For example, after *Dr. Eldredge* made the statement [in 1979] about the horse series being the best example of a lamentable imaginary story being presented as though it were literal truth, he then *contradicted himself...* On February 14, 1981 in California, he was on a network television program. The host asked him to comment on the creationist claim that there were *no examples of transitional forms to be found in the fossil record.* Dr. Eldredge turned to the horse series display at the American Museum and stated that it was the best available example of a transitional sequence." - *L.D. Sunderland (Darwin's Enigma (1988), p. 82)*

Why *Similar Structures* are NOT Evidence of Evolution

The study of *similarities* is the study of likenesses between various types of creatures. For example, both man and a number of other animals have livers, intestines, and appendixes. Therefore, **according to the evolutionary theory of similarities, they all descended from a *common ancestor*.** Evolutionists use the term *homology* to describe these similar structures, and consider them to be an important evidence of evolution.

If you compare a human arm with the front leg of an alligator or horse, or the flipper of a whale or a bat's skin-covered wing - you will find they all have a similar arrangement and number of bones.

Although *similarities* are considered by Darwinists to be an important evidence of evolution, keep reading and we'll find that the subject really proves *nothing* at all.

SIMILAR STRUCTURES... The proof that Darwinists really need is overwhelming *evidence of species change,* **and** *NOT similarity of structure or function.* Lacking that evidence, an attempt to prove the point by appearance is shallow at best. The big problem for evolutionists is that evolution is *not* occurring now, and the fossil record reveals it has *not* occurred in the past.

Yet there are many ways in which different kinds of plants are alike. The same holds true for animals. Since these similarities do exist, let us consider them briefly.

Physical similarities in plants, and in animals, can have two possible causes:

(1) **They either indicate that those creatures that are similar are closely related,** or (2) **they show that a single** *designer* **with** *immense intelligence, power, and ability made creatures with similar designs.* Evolutionists call these similarities, *"homologies."*

Stepping into a kitchen... you will find forks, knives, and spoons. Close examination will reveal that there are big spoons, little spoons, and even serving ladles, as well as five or six types of knives. Does this prove that the large spoons descended from the little spoons, or does it show that someone intelligent made them all? The spoons were made to hold liquids, and the knives were made to cut solids. Someone *designed* each of them to do a special work. They were produced by a planner and maker.

The above illustration focuses our attention on *purposeful design* and an *intelligent designer.* (1) There are similarities in the structure - the outward appearance - because of the purpose they must fulfill. (2) The spoons did *not* make themselves by accident, nor are they the result of a *chance arrangement of molecules.* They were *designed and manufactured by someone intelligent.* Even if they were made by machinery, someone very intelligent designed and made that machinery.

Whether it is similarities of spoons, similarities of eyes, or similarities of arms - the **answer is "Creation according to a common *design.*"** That is why Chryslers and Chevrolets are more alike than Jeeps and Yachts. Automobiles have many features in common because they were all designed to roll down highways, powered by engines. Sailboats are also very similar to one another because they were designed to travel by wind power over the surface of the water.

Turning our attention from man-made things to living organisms, it is equally obvious that similarity of structure follows *purposeful design* here as well. Neither *haphazard random activity* nor *accidents* can ever produce useful organs. *Intelligent planning* is *required.* INTELLIGENCE *cannot* be produced by *random, unguided NON*-INTELIGENCE.

DIFFERENT STRUCTURES... Not only do different animals have certain similar structures - they have *different* ones also! If they didn't, they would all look alike! So there are differences as well as similarities. For example, consider dogs and cats: There are a number of similarities between the cat and dog families. But look at all the differences! There are so many of them. Dogs are man's best friend. Cats are from the devil. Actually I like cats. It's my pastor who doesn't.

As we consider those differences, the idea of a common ancestry fades out - especially when there is *zero* evidence in the past *or* present that even one animal and plant type ever changes into another.

The *differences* emphasize the factor of a *common "designer,"* just as the *similarities* do, too. Examining these differences more closely, we find that each species, or basic type of plant or animal, has *unique qualities* that the others do *not* have. Yet **even those differences were** *purposefully designed.*

Amazing functional structures are also to be found in non-living things. For example, consider the exact specifications found in the orbiting of nuclear particles in the various elements. View the exquisite formations that various chemicals make as they crystallize. Each chemical always crystallizes in a certain way. Absolute evidence for *supernatural design*, certainly <u>not</u> *accidental evolution.* - *EvolutionFacts.com*

Pathetic Arguments for Evolution

Next, let's briefly look at a list of the *"strongest evidences of evolution," as presented in a number of evolution textbooks,* **found at EvolutionFacts.com** (except #35). **Please try to hold your laughter back until the end, many of these are pretty dog-gone funny (and stupid beyond belief).**

They say that EVOLUTION IS TRUE because... **1** - Aristotle taught evolution. **2** - Linnaeus classified plants and animals. **3** - Darwin wrote an influential book. **4** - Morgan studied fruit flies. **5** - Every living thing has chromosomes. **6** - People age as they become older. **7** - All living things have cells. **8** - All birds have feathers. **9** - Woodpeckers punch holes in trees. **10** - Birds breed in different climates. **11** - There are both light and dark moths. **12** - Some species have become extinct. **13** - Mendel discovered inheritance patterns. **14** - Coin tossing exemplifies evolution. **15** - DNA is the key to inheritance. **16** - Variants exist among people. **17** - Changes have taken place within species. **18** - Mutations produce new characteristics. **19** - Migration may cause evolution. **20** - Mating preferences may cause evolution. **21** - Predatory killing may cause evolution. **22** - Owls eat white mice first. **23** - Birds eat peppered moths. **24** - Different bears are different sizes. **25** - Teeth become smaller with age. **26** - Mutations produce sickle-cell anemia. **27** - A fish must have climbed out of water. **28** - Time should be able to produce evolution. **29** - Evolutionary *charts* prove long ages. **30** - Minks change color in winter. **31** - Stone tools have been found. **32** - Dinosaurs became extinct. **33** - Some earlier people lived in caves. **34** - Cave paintings have been found. And, finally, evolution is true because… **35** – Some people believe that Uncle Charlie was Right.

Yes friends, this pitiful list is basically all that the top evolutionists have been able to make up over the past 160 years. They have nothing to back up their invented and unimaginative stories and lies. All of the theories of evolution are a joke and a totally pathetic fairytale for all ages.

How to Beat ANY Atheist/Evolutionist at the... *Creation vs Evolution* Game

Be on Your Guard... as staunch evolutionists (not all, but rather the majority who are either agnostics, or the *new generation* of modern day atheists) will try to come down hard on you with *weak* evolution related arguments - that don't hold water. Actually *every one of them* *is weak and anemic,* they just don't believe or accept it. Or, they'll ask you silly questions about creation/evolution that they feel that you can't answer well... such as, *"Prove there is a God,"* or *"How did Noah fit all the animals on the ark?"* or *"Why does the Bible contain so much anti-scientific nonsense?"* Don't fall for their little traps.

Since Evolution is a Colossal LIE, basically just "Monkey Science" as I like to say... *usually* their related questions are *not* valid, at all, and in almost all cases certainly *not* the most important questions to ask. **Educated creationists can indeed answer the many off-the-wall questions asked by evolutionists, but the** *most important questions* **must be addressed by** *evolutionists - FIRST.*

The fun thing for creationists is that no matter how educated the evolutionist, they can't answer these questions intelligently or correctly unless they want to debunk their own theories. They are beat even before the battle begins. They don't stand a chance, never have, never will. I *almost* **feel sorry for them. Well, Almost.**

Now, if you're a staunch atheist, you're probably fuming mad at me right about now. Right? However, *maybe you should consider taking that anger out on the "theories" that have deceived you for a lifetime and robbed you of the truth.* Maybe it's your time to switch teams. It's never too late. God is waiting for YOU with open arms.

Back to the ways to conquer the atheist and evolutionary mindset... Many (maybe even *most*) times I will *not* answer their silly questions until they *first* answer my questions. I suggest you do likewise.

It usually works best if you *turn their questions back at them* saying something like... **"Let's prove the** *'natural'* (i.e. *the* **catatonic** *theory of evolution) before we try to prove the 'supernatural' (Intelligent Creation).* **So let's start with the most important questions first.** *You answer these five questions, and then I'd love to get to yours."*

Tough Questions to Start With

1) How exactly did the universe began from NOTHING, and can you *prove* it as a *fact*? (You can't say, "It started from a *tiny speck* of whatever." Where did that *speck* come from then? - If they have some bogus answer, then ask them, ***"Can you, or anyone for that matter, prove this is true?"***). "I just know it's true," is *not* a good answer, unless they're under 5 years old.

2) Exactly how did the *first* cell form? (The *second* cell? Was one a male and the other female who lived in the apartment next door, at the same time in history, so they could procreate?)

3) Where are the BILLIONS of *transitional* fossils that *must* be found in the fossil record for evolution to be true? All we ever find are *fully-formed* fossils from the bottom stratum all the way up to the top. **NO *in-between* species are found. Ever. Maybe they're just shy and hiding under a rock.** Pun intended.

4) *Prove* that even *one* species has evolved into a *different* kind (i.e. rat to horse, cat to monkey, rose to oak tree, etc). Hint: Darwin could *not,* nor can *any* top scientist today.

5) What is your strongest *evidence* for evolution?

Again, it makes no sense trying to answer their questions until they first answer these. **Don't worry, nobody can correctly answer any of these *most important questions*** (not even the top evolutionary scientists in the world)... As I have been saying from page 1, *the theory of evolution is a fairytale.* One of the biggest lies of all time.

IMPORTANT: I have found that you will find it much easier to... ask the evolutionists the questions *first,* **or give them this book,** *rather* **than letting them ask you the first questions.**

The reason being, when some skeptics ask you questions, creationists usually experience that the evolutionist is *not* listening to your answer. Many times they're trying to make you feel stupid in the process of proving that they're right. For the most part, they think the two weeks of teaching on evolution they got many years ago in school is all true and that's all they need. **Most know** *very* **little about either intelligent design or evolution - even most science teachers don't, so please don't just assume that they do.**

Since it's *naturalism,* the materialist (*evolutionist*) must *prove* their natural *(and comatose)* evolutionary theory if they are to be taken seriously. **Creationists do** *not* **need to prove the existence of God to the evolutionist, since He's...** *Super-Natural.* **By** *disproving* **natural evolution, in essence we** *prove* **that** *supernatural* **(creation by God) is true. There are** *no other* **options for the creation of all living things, be it animals or plants.** *Intelligent* **Creation Wins.**

More Questions to Ask Atheists and Evolutionists *(or Yourself, if UR1)*

Help the skeptics understand that virtually every aspect of their failing theories of evolution are misrepresentations, misunderstandings, unscientific, or flat out lies. How? By respectfully asking them important questions. Start with these...

Question... Evolutionists say that the universe was created from *nothing.* Please explain the *evidence* for this?

Question… Why do you believe that *"nothing"* created the universe, *rather than God?*

Question… What's your strongest *evidence* for any evolution?

Question… Please explain how the *Earth, Sun,* and the *Moon* came into existence, and then how they *continue* to stay where they are in space so life on Earth continues. (If you mention *gravity,* then explain how that came into existence, too, and how it knew it had to work closely with the planets).

Question… What is your biggest problem for the existence of God?

Question… Can you prove that God does not exist?

Question… There are 100 to 200 million fossils in museums around the world. This means there MUST have been multiple billions of *transitional* fossils found as well if evolution is true and creation is false. It's *impossible* for them to be hidden. Reptiles with half lungs and half gills aren't going to make it. *Think about it, half of either isn't going to work, such as birds with half wings, etc.* There are no (zero) real *transitional* fossils in the fossil record. So, my question to you is, If you think there are *transitional* fossils, *where* are they, exactly?

Question… Let's dig deeper on the Sun. How is the Sun the *exact size* and *temperature* it is, and why is it the *exact distance* from Earth it needs to be to sustain life? And, why does it *continue* to stay this same distance from Earth? Why is the gravitational force *just right* so as not to suck the Earth up into it? How did gravity come to be in the first place... all so we can sustain life? (The mathematical odds that *design-less chance* made any one of these happen is absolute *zero,* no matter how much *time* or *imagination* you have. If *any* of these things were *slightly* different, planet Earth would cease to exist, and so would we. This could *only* happen by a *supernatural* force).

Question... How *exactly* did life begin? (How does anybody know?)

Question... How does a human baby grow in roughly *9 months* when evolutionists say it took *millions of years* for humans to evolve? And, how do they *live in water* this entire time, but then once outside the womb, will die in a few minutes if put back into water?

Question... Evolution says *every* living species evolved from another. Give me just *one* example of a species that evolved into another kind of species. Like a reptile to a bird. (Don't feel bad, not even the top evolutionary scientists in the world have an answer or any evidence. Not for lack of trying or faking it for the past 160+ years).

Question... Where can we find the multiple BILLIONS of *intermediate/transitional* fossils that *must* be in existence, to prove evolution?

Question... How did a simple cell decide to become an acorn seed? Then, how did that acorn became an oak tree and not an apple tree, pear tree, monkey, or *anything else* for that matter? **Please tell me the exact *"mechanism of chance"* that produced this evolution** (not the: *what kind of mechanism,* such as natural selection, mutations, etc, - but rather *how it happened* and *happens* today producing *different* kinds of species, and *not* little changes within the same species which is simply *adaptation*).

Question... How did the blood get around the body before the veins and arteries were formed? Wouldn't they all have to be *fully-formed* from the beginning in order for any living animal to live more than a few minutes?

Question... How does a butterfly go from the *liquid* pupa stage to a beautiful *fully-formed butterfly* in 1 to 2 *WEEKS?*

Question... Why does a plant's roots decide to grow *downward* and it's stalk grow *upward?*

Question... So, most evolutionists say that monkey and human DNA are about 98% identical. First of all, it's actually been proven to be closer to 88% (and if you look even deeper, it's actually *much less* than that because most DNA tests have been greatly skewed to show the results evolutionary scientists want and need to fit their *anti-creation worldview.* More can be learned from *Dr Georgia Purdom* at AnswersInGenesis.org). It's said that even *mice* and humans share 90 percent of the same DNA. Humans and *chickens* are 65% the same, and even *bananas* share 50% of the same DNA as humans. Similar DNA simply means a *common creator,* certainly *not* evolution. However, you can throw out any statistics you want, but where are the *transitional fossils* between species? They are the *evidence* you need.

Question... Why do you believe we all just happen to exist by *random chance?*

Question... Even though there is *zero evidence* (and it's not for lack of trying to find it or falsify it)... you believe you evolved from monkeys who evolved from pigs (or whatever), who evolved from reptiles (or whatever), who evolved from fish (or whatever), who evolved from an amoeba (or whatever), who evolved from dirt and water (or whatever), who evolved from stardust... ***but you can't believe in an Intelligent Creator?*** How do you explain that logic?

Question... Do you believe what evolution teaches - that we are all related, meaning you are related to an acorn, an ear of corn, a sea slug, and a raccoon? Thus, we're all one big happy family.

Question... Explain how evolution can possibly work within the law of *Irreducible Complexity.*

Question... Were you aware that every fossil (billions of them) found in the lowest strata level (the Cambrian), is *fully-formed*, with *no* transitional fossils ever found *before* them? And, that other fossils that are exact look-a-likes to these Cambrian fossils, found many stratum levels up, and supposedly millions of years younger, are virtually *identical* to the lower level fossils with *NO transitional fossils* found in-between? Please explain how this could possibly happen, and why there aren't any *transitional* fossils, anywhere on Earth?

Question... Are you willing to give up a perfect eternity with God for the fairytale lies of evolution?

Question... If you can believe all the evolution gobbledygook, double-talk (Life starting from non-life, which has *no* scientific evidence to support it whatsoever), don't you think you could be open to consider *intelligent* creation by God? To explore that deeper start at CharlieWasWrong.com and more importantly, the Bible.

ATHEISTS KNOW GOD EXISTS
By Ray Comfort (Founder of LivingWaters.com)

Do Atheists hate God? Yes, they do. That's why they like to argue but they don't go to places online to spew hate against fairies, leprechauns, unicorns or purple spaghetti monsters. Why not?

How about Superman or Spiderman pages? They don't believe they are real, but they don't seek those pages on which to spew their hatred and ridicule. Again we ask, Why not? Because, of course, those do *not* exist. What is to hate?

God exists. He's real. Atheists know it, too. *Romans 1:18-20* tells us that Atheists are suppressing the truth. That is, they are acting in denial of reality. That's a delusion, which is a mental disorder.

Don't want to call it a mental disorder? Then call it this way, the only other choice: Atheists are *pretenders* and *liars*. And because **Atheists know God exists, they are without excuse come judgment day.** *(Romans 1:20-25, John 8:44)*

The Bible says in Romans 1:20-25...

"For the invisible things of him from the creation of the world are clearly seen, being understood by the things that are made, even his eternal power and Godhead; so that *they are without excuse:* Because that, when they knew God, they glorified him not as God, neither were thankful; but became vain in their imaginations, and their foolish heart was darkened. *Professing themselves to be wise, they became fools,* and changed the glory of the incorruptible God into an image made like to corruptible man, and to birds, and four-footed beasts, and creeping things. Wherefore God also gave them up to uncleanness through the lusts of their own hearts, to dishonor their own bodies between themselves: Who changed the truth of God into a lie, and worshipped and served the creature more than the Creator, who is blessed for ever."

Are these possibly the last days or years? **"Above all, you must understand that in the last days scoffers will come, scoffing and following their own evil desires. They will say, 'Where is this 'coming' he promised? Ever since our ancestors died, everything goes on as it has since the beginning of creation.' But they deliberately forget that long ago by God's word the heavens came into being and the Earth was formed out of water and by water. By these waters also the world of that time was deluged and destroyed."** - *2 Peter 3:3-6*

Almost 250 "D" Words Below That Can be Used to Address, Describe, and Help Destroy... the Evil Fable of Evolution.

Backstory: *I started using the words Destroy and Debunk one day as alternatives to words like Annihilate, Kill, Pulverize and Crush, to describe what will happen to the deceitful theory of evolution when it's really exposed... and over time it grew to almost 250 words that all start with the letter D. Some of the "D" words just address or describe evolution, others destroy and debunk it.*

Destroy, demolish, dismantle, debunk, defame, defeat, disassemble, deface, delete, devour, dissolve, devastate, demonic, despicable, deceit, deceitful, disgusting, dangerous, disturbing, disdain, delusional, denounce, devolve, discredit, deplore, decry, devitalize, disown, dispose, disorder, downsize, deprive, death, decimate, disturb, drain, despoil, demoralized, de populate, disparage, deceive, destruct, derogate, diminish, discomfit, diffuse, detonate, doom, digress, devil, daze, disjoin, disapprove, deport, defy, depraved, demoralize, doubt, dark, divisive, demon, disorganized, division, dissever, desert, divide, disesteem, deteriorate, die down, disconnect, dishearten, disperse, do in, disorient, disengage, dissociate, devalue, disapprove, deprecate, disband, distain, disdain, disregard, distasteful, dissimulate, disfavor, declaim, dumbfound, defile, dwindle, disaffirm, dishonor, decompose, devaluate, disloyal, decay, decline, disassociate, displace, downturn, dilapidated, dislodge, dispeople, desolate, deracinate, disentangle, diverse, dislikable, discrepant, dissimilar, disembowel, disruptive, desperate, disintegrate, distress, demobilize, decadence, dethrone, destruction, detach, dump, debauchery, decouple, diffused, deflate, dispersion, defecate, disgrace, discharge, dispute, deject, disquiet, disincline, disunited, decommission, disclaim, disappoint, disrupt, depress, disable, disavow, dispirit, die, death trap, divarication, defunct, divorce, dissect, dampen, decrease, debilitate, downgrade, desiccate, dilemma, deaden, dismiss, dynamite, declare, depreciate, depopulate, detestable, disallow, disthronize, detour, deplete, disarm, disillusion, discompose, dry out, degrade, dry up, disprove, dehydrate, debase, demonetize, dilapidated, drop, deny, dash, define, distant, disappear, deconstruct, demur, dislocate, deride, divest, damnable, deranged, degenerate, dispel, disrelish, distressing, deprivation, disagree, disconfirm, discombobulate, dizzy, dingy, dung, dinky, dorky, dire, dreadful, daunt, demean, distressing, dispraise, discount, drastic, desecrate, discountenance, disconcert, deceitful, disdainful, disparate, discordant, discreditable, daring, disreputable, dirty, deprogram, dodge, duped, deescalate, disabled, diverge, disapprobation, dim, desist, divergent, distrait, dehumanize, discourage, dilute, displease, dishonest, damage, deform, dichotomize, disrepute, dismember, distraught, disfigure, deify, dissipate, destitute, dissolute, dishonorable, dematerialize, derail, *and one more time...* DESTROY.

Once I got started, I couldn't stop myself. :-)

Ambassadors Wanted

Our goal is to get this book into virtually every high school and college in North America in the next 3 years. Then, worldwide.

Meaning, we need <u>YOU</u> to order several books and get them to youth pastors in your city and state.

The pastors and/or student leaders can organize study groups before and after school, even on lunch breaks.

Or, do it on the Internet, or in homes when schools are closed. Make sure to abide by *social distancing* laws if they are being enforced.

Join the Movement
<u>CharlieWasWrong.com</u>

Evolution is DEAD!!
JESUS is ALIVE!!!

Evidence for the Bible

The Bible is the best selling book of all time. By a long shot. Bible means *Biblia* in latin. Meaning a "**collection of books.**" It can be *historically verified* through **archeology**, **manuscripts**, and *many* **eyewitnesses** (Christian *and* non-Christian).

Several first-century *non-Christian* historians and writers confirm the life, and death on the cross of Jesus: *Cornelius Tacitus, Lucian of Samosata, Flavius Josephus, Suetonius, Pliny the Younger, Thallus, Phlegon, Mara Bar-Serapion,* and references in the *Talmud* and other *Jewish writings.* - *Encyclopedia Britannica*

WE CAN BELIEVE... what the Bible authors originally wrote. Let's look at some of the *evidence.*

1) About **24,000** *handwritten* **copies of the 27 original New Testament books** (About 6,000 in Greek. Many full copies and some fragments) were written **as early as 40AD to 50AD** with the oldest NT manuscripts discovered around 125AD. (By comparison, Homer's "Iliad" comes in second with only 643 copies discovered, the earliest found being about 500 years after his death).

2) The **Bible's 66 books were written by 40 authors** from **all walks of life,** over **1,500 years** in **3 languages on 3 continents,** yet it reads as *one* **story.** The amazing continuity of the Bible is due to the fact that ultimately it has only *one* author (GOD) directing human writers.

3) Instead of always being the *hero,* many of the writers wrote about how *flawed* they were in many different circumstances. If the Bible was a bunch of made up stories, don't you think the writers would be the heroes, not the *zeros*?

4) Several *non-Christian* historians wrote about the existence of *Jesus Christ*.

5) Except John (and Judas of course), **all the apostles died horrible, excruciating deaths for a *reason*, that being *JESUS*.** Even Peter, who denied knowing Jesus 3 times the night before the resurrection, was *crucified upside-down* years later for his love and ministry for Him.

MORE EVIDENCE... The Jewish scribes solidified the following process for creating copies of the Torah and eventually other books in the Old Testament.

1. They could only use **clean animal skins**, both to write on, and even to bind manuscripts.

2. Each column of writing could have no less than forty-eight, and no more than sixty lines.

3. The **ink must be black**, and of a special recipe.

4. They must *verbalize each word aloud* while they were writing.

5. They must wipe the pen and wash their entire bodies before writing the word *Jehovah*, every time they wrote it.

6. There must be a **review within *thirty days*,** and if as many as three pages required corrections, the entire manuscript had to be **redone**.

7. The letters, words, and paragraphs had to be counted, and **the document became invalid if two letters touched each other.** The middle paragraph, word and letter must correspond to those of the original document.

8. The documents could be **stored only in** *sacred* places (synagogues, and the like).

9. As *no* **document containing God's Word could be** *destroyed,* they were stored, or buried, in a genizah - a Hebrew term meaning *hiding place.* These were usually kept in a *synagogue* or sometimes in a *Jewish cemetery.*

Even though the evidence and writing of the Bible happened *thousands of years ago...* **the** *important thing* **was that the** *original text* **was** *written a "short time"* **after the** *actual events.*

There is **better documentation** for the *life of Jesus* than for all of the major figures in history, *combined.* Biographies were written about Alexander the Great 400 *years* after his death. On the other hand, **records for the life of Jesus were written with the** *first generation,* **while** *many of the eyewitnesses were still alive to dispute them* **if they were incorrect or made up. If they were made up, we would also have documentation of that preserved as well, but we do not.**

The Bible *accurately documents* **353 prophecies** (written hundreds to thousands of years *before* the birth of Jesus) **were** *all* **fulfilled by Jesus,** *exactly* **as prophesied.** (The probability of only "8" of these 353 prophecies being fulfilled is 1 in 10_{17} power, or 1 in One Hundred **Quadrillion**!! In other words... **IMPOSSIBLE!!!**

The Bible was written *thru men by God.* 2 Peter 1:21 says, "Holy men of God spoke as they were moved by the **Holy Spirit.**"

Modern Science was birthed in the 17th century because of a belief in an *unchanging* **God of** *order, purpose* **and** *consistency.* An example is... when scientists thought that the Earth was flat, the Bible says in Isaiah 40:22, "It is he (God) who sits above the *circle* of the

Earth." There are *several* more science related examples in the Bible. **The Bible's remarkable *continuity* from beginning to end is more *evidence* of its *divine origin*.**

We should believe the Bible: It is **Reliable**. It is **Proven**. It is **Truth**.

"It is impossible to rightly govern a nation without *God and the Bible*." *- George Washington*

*The above information I got from a local life group my wife and I wrote and lead, but I don't have the citings from where it came from. I know it probably came from one of two of the sources I have listed at the beginning of this book. Please visit all of them online and buy their excellent books and videos.

Science Facts (in the Bible)
(from Ray Comfort's book, *Science Facts in the Bible*)

EARTH'S FREE FLOAT IN SPACE... Job 26:7 (written 3,500 years ago): *"He stretches out the north over empty space; He hangs the Earth on nothing."*

The Bible proclaims that the Earth freely floats in space. Some in ancient times thought that the Earth sat on a large animal. We now know that the Earth free floats in space.

THE EARTH IS ROUND... Isaiah 40:22 (written 2,800 years ago): *"It is He who sits above the **circle** of the Earth."*

The Bible informs us that the Earth is *round*. Though it once was commonly believed the Earth was flat, **it was the *Scriptures that inspired Christopher Columbus* to sail around the world.** He wrote: "It was the Lord who put it into my mind... There is no question the inspiration was from the *Holy Spirit* because He comforted me with rays of marvelous illumination from the Holy Scriptures…" (from his diary, in reference to his discovery of "the New World").

SECOND LAW OF THERMODYNAMICS… Psalm 102:25-26: *"Of old You founded the Earth, and the heavens are the work of Your hands. Even they will perish, but You endure; and all of them will wear out like a garment."*

The Bible tells us *three times* that the Earth is *wearing out* like a garment. This is what the *Second Law of Thermodynamics* (the Law of *Increasing Entropy*) states: that in all physical processes, every ordered system over time tends to become more disordered. **Everything is running down and wearing out as energy is becoming less and less available for use. That means the universe will eventually *wear out* something that *wasn't discovered by science* until fairly recently.**

THE ORIGIN OF SEXES… Matthew 19:4: *"He who made them at the beginning 'made them male and female.'"*

Almost all forms of complex life have both male and female - horses, dogs, humans, fish, moths, monkeys, elephants, birds, etc. **The male needs the female to reproduce, and the female needs the male to reproduce.** *One cannot carry on life without the other.* But if evolution were true, which then came first according to the dead theory?

If a male came into being before a female, how did the male of each species reproduce without females? How is it possible that a male and a female each spontaneously came into being, yet they have complex, complementary reproductive systems? **If each sex was able to reproduce without the other,** *why* **(and** *how***) would they have developed a** *reproductive system* **that** *requires* **both sexes in order for the species to** *survive?*

ASTOUNDING MATHEMATICAL ODDS... The Old Testament prophets declared, among many other things, that *the Messiah would be born in Bethlehem* (Micah 5:2) to a *virgin* (Isaiah 7:14), be *betrayed for thirty pieces of silver* (Zechariah 11:12,13), *die by crucifixion* (Psalm 22), and be *buried in a rich man's tomb* (Isaiah 53:9). There was only one person who fits all of the messianic prophecies of the Old Testament, who publicly performed countless miracles, made the crippled walk and the blind see, resurrected the dead, taught the most profound words ever uttered, and then died for the sins of the people, all before AD 70: *Jesus of Nazareth,* the Son of Mary.

Couldn't Jesus have "accidentally" fulfilled all the dozens of prophecies? NO. **The scientific probability that any one person could fulfill just *eight* of these prophecies is 1 in 10_{17}.**

Now let's try to imagine the likelihood of that. If we took that number of silver dollars (100,000,000,000,000,000), drew a black X on only one, and laid them over the state of Texas, they would cover the entire state two feet deep. Now *blindfold* a man and tell him to travel as far as he wishes and then pick up only one silver dollar, and it must be the marked one. What chance would he have of picking up the right one?

It would be exactly the same odds that just *eight* of the messianic prophecies would all come true in any one person - yet they *all* came true in Christ (adapted from *Science Speaks* by Peter Stoner).

There are over *three hundred* prophecies that tell of the ancestry, birth, life, ministry, death, resurrection, and ascension of *Jesus of Nazareth. **All have been literally fulfilled to the smallest detail.***

The Universe is Expanding... The Bible mentions several times about the *Expanding of the Heavens (Universe)*. I.E. *Isaiah 40:22.* This was written 3 THOUSAND years *before* the Hubble telescope discovered that the universe is expanding. *- Much more can be found at LivingWaters.com*

SURVEY SAYS... The vast majority of **College Students and Young Adults** - *who went to church as teenagers,* but who did *not* learn the *Biblical Creation Story* growing up are *not* going to church anymore. However, those who *did* learn about God's Creation, are almost all going to church. Please share this book and other creation related information with everybody you know from kids to adults, especially teenagers and youth pastors.

"God shall wipe away all tears from their eyes. There shall be no more death. Neither shall there be any more sorrow, nor crying nor pain, for the former things have passed away." *- Rev. 21:4*

Why is it So Important to Believe That *Biblical Creation* is True?

The Bible is *never changing!* Science on the other hand *changes all the time* to try to fit the facts that the Bible got right from the beginning.

1) **The first verse of the Bible identifies who God is; He is the *Creator, not* an idea, *not* a moral standard.** All scripture agrees and supports that point. From beginning to end, Genesis to Revelation, the concept of an actual, historical creation identifies our CREATOR.

2) **The second point is that creation *identifies* MAN.** ("Man" referring to male *and* female) ***Not* just a stack of chemicals that arrived by *random natural processes*. Man was created in the *image of God* -** body, mind, and spirit, and has *purpose.*

3) **The third point is that** *creation* **is the** *basis of the GOSPEL.* God created man. Man rebelled against God. God has a plan of *salvation.*

The Bible has been *proven* to be historically true. You can and *should trust* in it.

"I believe *the Bible is the best gift God has given to man.* **All the good from the Savior of the world is communicated to us through this book."** - *Abraham Lincoln*

Is ALL Scripture *God Breathed?*

"All Scripture is God-breathed and is useful for teaching, rebuking, correcting and training in righteousness, so that the servant of God may be thoroughly equipped for every good work." - *2 Tim. 3:16-17*

Keep in mind that the *entire* **Bible is considered** *Scripture, from God to us, not* **just parts of the Bible.** (It's his *love letter to us* so to speak). This verse does *not* say, *"SOME* scripture is God-Breathed..."

This verse is either true or it's not. Even though the New Testament was written after the death and resurrection of Jesus, if it was inspired by God Himself (written by God through men), don't you think that ALL of it, the Old *and* New Testament are GOD-BREATHED?

If it were not, then I would propose it could easily contain a hodgepodge of *all different kinds of writing within it* - due to the *ego of man.* What New Testament version (NIV, ESV, KJV, NKJV, NRSV, NASB, etc) would be *the* correct one as they seemingly could all be vastly different?

The Bible says that the guys who wrote the letters that became the New Testament were *writing as GOD himself directed and inspired them to write.* To believe differently opens up a *can of worms* (Note: **Paul refers to *Luke's* New Testament gospel writing as "*Scripture.*"** Paul writes… **"For *SCRIPTURE* says, 'The worker deserves his wages.'"** *- 1 Tim 5:18).*

And, **Peter** *also* **refers to *Paul's* New Testament writings as Scripture in 2 Peter 3:15-16.** These verses show us that the idea of adding new Scripture, in addition to the Old Testament, was already happening in early church years. It also seems to tell us that Christians were *expecting more Scripture to be added.*

Digging a little deeper yet, from *TrueLife.org.* Paul writes in…
1 Corinthians 14:37-38, "If anyone thinks that he is a prophet, or spiritual, he should acknowledge that the things I am writing to you are a command of the Lord. If anyone does not recognize this, he is not recognized." **In this passage, Paul directly addresses his own writings and calls them a *"command of the Lord."*** This is nearly a direct claim to divine inspiration. This idea of a command of the Lord is found throughout the Old Testament as a claim to direct revelation from God. For example, God gave commandments to the people of Israel through Moses (e.g. Exodus 35:4 Numbers 10:13 etc.), who spoke to the Lord "face to face, as a man speaks to his friend" (Exodus 33:11). Furthermore, **Paul even declares that if anyone does *not* acknowledge that his words are a command of the Lord, "he is *not* recognized." This shows Paul's confidence in the authority of his words.** It is also similar to the warnings that God gave to the Israelites if they failed to acknowledge and follow his commandments through Moses (Deuteronomy 11:26-28).

Paul claims divine authority for his words in other letters as well. 1 Thessalonians 2:13 says, "And we also thank God constantly for this, that **when you received the word of God, which you heard from us,**

you accepted it not as the word of men but as what it really is, the *word of God,* **which is at work in you believers."** Here Paul explicitly refers to the words that he and the apostles were preaching as the "Word of God." Paul appears to be claiming the same inspiration as the prophets of the Old Testament, who would begin with "thus says the Lord," or "the Word of the Lord."

The writer of the book of Revelation, John, makes an express claim to *divine* inspiration. At the beginning of the book, John writes that what follows is a revelation that he received from God through an angel. He then writes, "Blessed is the one who reads aloud the words of this prophecy, and blessed are those who hear, and keep what is written in it, for the time is near" (Revelation 1:3). John commands the reading of this book out loud and the keeping of what is written in it.

By doing this he equates his writing to the law of the Old Testament. God commanded the Israelites to read the law in the hearing of all the people, so that they could keep what was written in it (Deuteronomy 31:11-13).

1 John gives us another clear assertion of divine inspiration. The author of this letter writes, "We are from God. Whoever knows God listens to us, whoever is *not* from God does *not* listen to us."

(1 John 4:6) Here the writer claims to be from God, indicating that he believed his teaching to be *divinely* inspired. Moreover, he claims that **everyone who knows God listens to his words, and anyone who is** *not* **from God does** *not.* This assertion is very similar to what Paul said in 1 Corinthians 14:37-38. It is also similar to Jesus' words to the Jewish leaders in John 8. "'Why do you not understand what I say? It is because you cannot bear to hear my word. You are of your father the devil, and your will is to do your father's desires'" (John 8:43-44). **By stating that his words are heard by those who know God and**

unheard by those who do not, the author of this letter is making a claim to *divine inspiration.*

There are many more instances of New Testament authors making direct or indirect claims to divine inspiration, which we do not have space to discuss here. A more in-depth inquiry would reveal that the authors paralleled Old Testament books, claimed the authority of apostolic tradition, warned of imposters, and more. **We have shown, however, that at least some of the writers of the New Testament considered themselves to be writing authoritative Scripture. This is enough to disprove the view that the authors of the New Testament did not regard their own writings as** *Scripture.*

SIDE NOTE: Don't get me wrong, we're *not* claiming any scripture is above others as **they are ALL directly from God, and all ultimately important. The Bible is God's WORD** for us to choose to follow, or not. All of it - not what we pick and choose to believe as inspired scripture or not. We *can't* say that only Jesus' words in the New Testament, or the Old Testament, are scripture, but some parts are not necessarily scripture, as it *ALL comes from God* and was written thru men. He does *not* force us to believe scripture, or even in Him, but rather gives us that decision to make ourselves.

Now, that's a *loving God.* **It's** *your* **choice to choose Him, or deny Him and follow the teachings of** *hopeless random chance evolutionary creation.* No small decision. Speaking of *decisions,* let's continue with some *Jesus talk, shall we?* Is He who He claimed to be, the Son of God *and* God in the flesh? The *only* way to eternal life? Or was He a liar or maybe a lunatic? The cool thing is that *YOU get to make that decision for yourself.* Nobody's twisting your arm, not even God, even though He desperately wants YOU to choose HIM. He already chose you, now it's your turn my non-believing friends.

"He who thinks half-heartedly
will *not* believe in God; but
he who really thinks *has*
to believe in God."

- *Sir Isaac Newton*

The JESUS Factor

"For it is by *grace* you have been saved, through *faith* - and this is *not* from yourselves, it is the *gift of God*, *not* by works, so that no one can boast." - *Ephesians 2:8-9*

The real topic of this book is not nearly as much about the past or present... but actually about **where you and your loved ones are going to spend your *ETERNITY*.**

Oh yeah, there's an *eternity* after this life and I hope by now you're open to hear the good news. The news about *Jesus Christ* and how He can usher you and your loved ones into it. See, I had to go all *Jesus Freak* on you now. :-)

Contrary to popular belief among skeptics, being a *Jesus* follower (Also called a *born-again Christian*) is actually very AMAZING!! And, the vast majority of the *billions* of us worldwide who claim Jesus as our Lord and Savior, are *not* weirdos as portrayed by some of the liberal public and Hollywood. Are there a few kooks? Sure there are. However, most are actually rather normal as you probably know if you're not one of us (yet), we just march to the beat of a different drum - that being *Jesus*.

Are Your Bags Packed? Christians have an incredible *hope* that non-believers don't have, and that is of a perfect life someday in *eternity,* which is actually on a *new Earth* as the Bible talks about in the 21st chapter of the book of Revelation. Check it out, it's *really* cool. *Heaven on Earth* so to speak. And, don't tell me you can't believe in this if you buy into the barren evolution theory now.

If you're *not* a *Jesus person* already, I hope you're now open and interested to learn why you *definitely* should be. Why it's more real

than life itself (as you now know it). And, whether you believe it or not, why your *eternity* depends on it.

If it doesn't make sense to you now, that's ok. In fact, the Bible in 1 Corin. 2:14 tells us why it *doesn't* make sense to *non*-believers.

In other words, if you have *not yet* made Jesus your Lord and Savior, the stuff you just read probably sounds pretty crazy.

After you make Jesus the Lord and Savior of your life (prayer to do this follows below), *you'll be amazed how the Bible comes to life and opens your eyes.* Don't freak out now, it's not as weird as skeptics make it out to be. It's actually a *very* cool, eye opening, life changing experience.

"Always be prepared to give an answer to everyone who asks you to give the reason for the hope that you have. But do this with gentleness and respect." *- 1 Peter 3:15*

The Ways We Know Jesus Actually Existed Are...

1) THE **BIBLE**.

2) **Secular historians** such as: Josephus, Tacitus, Thomas Arnold, Pliny the Younger, among others wrote about Jesus Christ.

3) **Rival religions** such as: *Jews, Muslims, Buddhists, Mormons,* and many others talk about Jesus as if indeed He lived. *Think about it,* it would be more beneficial for these other religions to say He did *not* exist, but they do *not* and in fact *cannot*.

4) Look at a *calendar.* Now look at your *checkbook.* What *dates* do you see? Even our money has a date on it. They correspond to the life of *Jesus Christ.* Need I say more?

The culture of today has reduced Jesus to a *nice guy*, a *good teacher,* and a *great prophet.* Jesus *never* made those claims. Rather, **Jesus claimed to be** the *Son of God,* the *Messiah*, the *King of Kings* and *Lord of Lords*, the *First and the Last,* the *Alpha and Omega*, the *Bread of Life,* the *Savior of the World*, and even *God* in the flesh.

If He isn't what He claimed... He was either the greatest *liar* in the world, or a *lunatic* out of his mind. This *doesn't* match up with being a great prophet, good teacher or nice guy.

If Jesus was a *fraud*, He could have easily been locked up in prison. He *didn't* have to be *crucified* if He was simply a fraud (a liar or lunatic). In fact, why would they crucify some *big liar* or *looney tune?* They wouldn't.

However... **Jesus was seen as a *huge threat* to non-believers.**

Jesus taught *differently* than other spiritual leaders. Most spiritual leaders *point away* from *themselves* saying, "Don't look at me, look at God." Jesus' teaching was pointed to *himself* much of the time. He said things like this about himself... *I am the bread of life. If the son sets you free, you are free indeed. Come to me and I'll give you rest. If you receive me, you receive the Father. If you see me, you see the Father. I and the Father are one.*

Jesus forgave sins. Only God could forgive sins, thus He was essentially saying to the people that He was *God* or equal to God. That was considered blasphemy in non-believer's eyes.

JESUS SAID... **"The thief (Satan) comes only to steal and kill and destroy. I came that they may have life and have it abundantly."** - *John 10:10*

Do you think maybe Satan uses the seductive, lying theories of evolution to steal, kill and destroy... to pull people away from God? The Bible says he does. What say you?

Did Jesus Rise From the Dead? (Is He Alive Today?)

Well, for starters, maybe you can relate to this... our *checkbook* and *calendar* dates - relate to the life of a real person, *Jesus Christ,* who lived among us about 2,000 years ago. So, we know for *sure* that He was a real person, but is there more solid *evidence* that He actually rose from the dead and is the Savior of the world today? Don't take my word for it, let's explore a little.

"You believe that there is one God. Good! Even the demons believe that, and shudder." - *James 2:19*

So, maybe you do believe in God. The bigger question is, do you believe Jesus actually *rose from the grave,* like the Bible says? Let's explore the *evidence.*

1) **Precautions of the Romans...** There were between 10 and 30 Roman soldiers on duty over the tomb, so He would *not* have been able to be stolen from the tomb by His disciples.

2) **Would the disciples be *martyred* for a liar or lunatic?** 10 of the original 12 were brutally killed for their belief in Jesus. Paul was *beheaded.* Peter *crucified upside down.* James was *stoned* and *clubbed.*

3) **Jesus appeared to *500 people*** for 40 days after the resurrection.

4) ***Secular* (and Christian) history confirms** this proof.

5) **Missing body of Jesus.** His body was *not* later found in the grave. The guards would *not* have allowed anybody to take Jesus, as they would be killed for allowing this to happen.

"The Good News borne by our risen Messiah who chose not one race, who chose not one country, who chose not one language, who chose not one tribe, *who chose all of humankind!*"
- *Nelson Mandela*

If Christ *didn't* rise from the dead, then Christianity is a *lie*.

Is Jesus the *Only* Way to Heaven?

Short answer... **YES!!** So, maybe you think that's pretty narrow-minded. If you *don't* agree, hopefully you'll soon understand why Jesus claimed to be the ONLY way.

He gives you the choice to accept or reject this claim. No other options. He doesn't force us, but rather gives us *free will* to make Him the Lord and Savor of our lives, or to reject Him.

Of course you have heard this famous Bible verse...

"For God so loved the world, that He gave his one and only Son, that *whoever* believes in Him will *not* perish but have *eternal* life."
- *John 3:16*

Here's another Bible verse that you may have not heard, it's words from Jesus...

"I am the way and the truth and the life. No one comes to the Father except through me." - *John 14:6*

He's saying that NOBODY gets to Heaven except through HIM. You can except that or not.

I know that many of you are *not* remotely open to the possibility of Jesus being the *only* way. Not only atheists and agnostics, but those who are of other non-Christian faiths, and even some of you who are now even creationists. Some of you are still *not* open to *intelligent design* by God, just *yet.* I get that.

Even though growing and being confirmed a Lutheran, as a kid, I wasn't a true *Jesus* person until I was in my 20's. It did *not* happen when I was *confirmed* in 8th grade, as I have no clue what that was all about. I was a clueless 14 year old kid. Like many teenagers of my era, I got into a lot of *bad stuff* back in those days. In high school I remember making fun of my *Jesus Freak* friend, Jeff Peterson, even one time in front of a hundred kids in the lunchroom, when I stood on top of a lunch table, pointed to him and shouted, *"Look everybody... it's Preacher Pete the Jesus Freak."* He was a great example of Jesus for me. He still is to this day. Now we're both Jesus Freaks and not afraid to tell the world. :-)

Then, one day a few years later, I prayed and asked Jesus to be my *Lord and Savior.* He became very real to me. Life took on new meaning. My eyes were opened so to speak. I was excited like never before and I couldn't stop myself from telling people about Him.

That was long ago. And, still today, I'm more and more in love with Jesus each passing day. That's the reason for this book. I hope and

pray it's words are reaching YOU. I pray you give *your* life to Him.
If you're *not* a Christian yet, do you think you're reading this book right now by - *random chance*? No way. Instead, why not be open to think it was by *divine appointment*, that God wants you to spend eternity with Him and *NOW* is *your* time to *surrender your life to HIM*. The Bible clearly says that, He will accept you *just as you are* right now, no matter what you did in the past.

My "Jesus Journey" has been incredible. I'm far from perfect and have had my fair share of problems over the years, just like you and every other human. I'm *not* any better than you or anyone else, I'm just *better off* than those who haven't made Jesus their Lord and Savior. You may doubt that and will never know what I mean… that is until you make a decision to make Him *your* Lord and Savior. Just make sure you're *sincere* and you *mean it* with everything in you when you reach out to Him in prayer. *Making Jesus your Lord and Savior is the most important thing you will ever do in your life!!!*

Does What WE *Think* Matter, OR... Does What GOD *Says* Matter Most?

Let's quickly explore what the Bible says about that and how we can spend a *perfect eternity* with the Creator of the World.

In the gospel of John (14:6) Jesus says, *"I am the way, the truth and the life… no one comes to the Father except through me."*

In other words, **Jesus himself said that there is NO other way to spend eternity with God. He *doesn't* leave it open for interpretation. It's <u>NOT</u> - *Buddha, Allah, Muhammad, Confucius, Krishna, the Dalai Lama, or any New Age doctrine or guru.* It's Jesus and <u>ONLY</u> JESUS!! Accept it or deny it, it's *your* choice.**

Many will say that this is *very narrow minded.* But is it really? Hold back any judgement for a bit, please.

If Jesus is *not* really who He said He is - the son of God *and* God in the flesh, then it *is* indeed very narrow minded. **HOWEVER... if Jesus actually *is* God incarnate** *(God in the flesh)***, and He is the Son of God,** like He claimed to be - **then it's *not* narrow minded at all, is it? If He is who He claimed to be... then He truly is God.**

The big question is... are you listening to *Him,* **or, are you listening to the** *hopeless man made doctrine of evolution?*

His disciple Peter said the same thing in Act 4:12, "*Salvation* **is found in** *no one else,* **for there is** *no other name under heaven* **given to mankind by which we must be** *saved."*

Next, this is so cool. It refers to JESUS as the "WORD."

"In the *beginning* **was the** *Word,* **and the** *Word* **'was with' God, and the** *Word* **'was' God. He was with God in the beginning. Through Him** *all* **things were made; without Him** *nothing* **was made that has been made."** *- John 1:1-3*

If the Bible was merely written by men only, and it's *not* the *infallible* and *inerrant* word of God, *co-authored* with men by the Holy Spirit as it plainly says... then these words of Jesus in John (Chapter 1) would be narrow minded.

However, **as you study more and open yourself up to God's word (the Bible), you'll learn that Jesus is indeed who He said He was/is. The Bible is indeed the inerrant and infallible... WORD OF GOD!!**

It's His *love letter* to us. Our "life" *instruction manual.* The more you read His amazing book, the more you'll feel and understand His

AMAZING LOVE FOR YOU!! Yes, God is REAL my friend, and He wants YOU!

Jesus said, "**I am the resurrection and the life. The one who believes in me will live, even though they die."** - *John 11:25*

Jesus is King *(Not Elvis)*

"I am *not* **the King. Jesus Christ is the King. I'm just an entertainer."** *- Elvis Presley*

And, now in more recent times, rapper *Kayne West* has committed his life to Jesus and raps about Him in his album JESUS IS KING. *Justin Bieber* has also said he's a Jesus follower. Several other celebrities, musicians and sports figures claim to follow Jesus, too. These people are no better than you or me, however, they are in the limelight and hopefully will humbly walk the Jesus walk and boldly share His love and salvation message with their millions of fans. Pray for them.

On a lesser note, we could also say that *"Evidence" is King,* too. As we have discussed throughout this book, *random chance evolution without actual evidence, is powerless and meaningless.*

Always keep in mind, **anybody can** *say* **anything they want, either for or against** *evolution* (or *intelligent design* for that matter), and even sound very convincing, authoritative and intelligent, but **that** *doesn't* **mean they are correct.**

Between the two, *Evolution* **and** *Intelligent Design…* **one is** *true,* **the other is a** *lie.* **They are opposites. They** *can't* **both be true.**

Many *non-creationists* will give misinformation and even say blatant lies *against* intelligent design creation. Some innocently out of ignorance of the actual facts, and some intentionally out of defiance, or maybe to sell books and speaking engagements.

Warning: Don't be deceived just because they sound convincing. Ask them to *prove* what they are claiming to be the truth, like… *"Hey guy who says we all come from stardust (or a monkey), you can say anything you want but **I need to see the evidence.** Show me the proof, man."* If you're now asking this same question to the creationist about proof for God, please *reread this book* because you missed a LOT of good stuff.

The *Roman Road* to Salvation

We all SIN. We all DIE. (Unless the *rapture* happens in our lifetime and it may. This is talked about in the book of Revelation and other books in the Bible where God talks about taking His believers to meet Jesus in the clouds, near the end of the world. Another story for another book). There is *HOPE.* His name is *JESUS.* Here are 4 verses from the book of Romans in the Bible that tell the story in short form.

"All have sinned, and fall short of the glory of God."
- Romans 3:8

"God demonstrates his own love for us in this: While we were still sinners, Christ died for us." *- Romans 5:8*

"The wages of sin is death, but the gift of God is eternal life in Christ Jesus our Lord." *- Romans 6:23*

"If you declare with your mouth, *Jesus is Lord,* and believe in your heart that God raised him from the dead, you will be saved."
- Romans 10:9

C.S. Lewis *(Lord, Liar, Lunatic)*

C. S. Lewis was a professor at both Oxford and Cambridge University. He is best known for *The Chronicles of Narnia*, and *Mere Christianity*. Lewis and fellow novelist **J. R. R. Tolkien** (*Lord of the Rings* author) were close friends. They were both Christians and professors at Oxford as well.

J.R.R. Tolkien led his atheist friend C.S. Lewis toward faith in Jesus. The *Lord, Liar, Lunatic* theory has become famous over the years.

Lewis offers these 3 choices about Jesus.

"I am trying here to prevent anyone saying the really foolish thing that people often say about Him: 'I'm ready to accept Jesus as a great moral teacher, but I don't accept his claim to be God.' That is the one thing we must not say. A man who was merely a man and said the sort of things Jesus said would not be a great moral teacher. He would either be a lunatic - on the level with the man who says he is a poached egg - or else he would be the Devil of Hell. You must make your choice. **Either this man Jesus was, and is, the Son of God, or else a madman or something worse.** You can shut him up for a fool, **you can spit at him and kill him as a demon or you can fall at his feet and call him Lord and God,** but let us not come with any patronizing nonsense about his being a great human teacher. He has not left that open to us. He did not intend to." *- C.S. Lewis*

Let's Go a Bit Deeper Into JESUS

As mentioned already, Jesus did indeed live on Earth as is supported by loads of documentation (both secular and the Bible). However, **the more important questions may be,** *did He rise from the grave* **and** *is He alive today? Have you committed your life to Him?*

An Evolutionary Fairytale

Now, you don't even believe any of this if you're an atheist who doesn't even believe there's a Creator, I get that. Or, if you believe in God but don't believe that Jesus is the *only* way to heaven and eternal life, you *must* then believe that He was either a liar or a lunatic when He said things such as in John 14:6 - **"I am the way, the truth and the life. *No one* comes to the Father except through *me."*** We really have *no* other choices.

Which brings us to – *JESUS IS LORD...* Multiple millions go as far as *believing* that Jesus is the *Lord and Savior* of the *world at large,* but have *not* asked Him into their *own* life, as the Bible describes. Does simply *believing* this give you eternal life? According to the *Bible,* the short answer is, NO!!

I grew up in a mainline denominal church yet did *not* have a *personal* relationship with Jesus until I was in my early twenties. I can't ever remember being taught what the Bible actually says about *salvation* (how to have eternal life). They may have, but I missed it somehow. Yeah, I know this all sounds freaky to many of you. It's really not all that weird, that's just how the evolution agenda teaches us to think and react towards Christians sold out to Jesus.

OK, let's talk about the *elephant in the room...* Many people *believe* Jesus is the son of God, the Savior, and even God himself in the flesh as the Bible clearly teaches, *BUT,* then are taught *incorrect "man-made" concepts* such as - doing **"good works"** and/or **"baby baptism"** give us *Eternal Life... but* **this is *NOT* at all what the Bible says.** It's a *man-made deception,* **don't buy into these man-made "ways to heaven."** It is clearly *not* what the Bible teaches. If you think otherwise, please show me undeniable *proof* in the Bible.

All 27 baptisms in the Bible were with *adults,* none were with babies. And, don't get me wrong, doing *good works* (James 3:13, etc) is what God wants, even commands us to do, however, He also says they are

NOT what gives us eternal life. **"By *GRACE* you have been saved through *FAITH.* And this is *not* your own doing; it is the gift of GOD, *not a result of works,* so that no one may boast."** - Eph. 2:8-9

So, we're left with the *choice* of believing He is Lord and Savior of the **world at large,** OR, we ask Him to be **our** Savior and Lord. There is a BIG difference. A little confusing? Actually it's simple, and he tells us just how to do that *(spend eternity with Him)* in the Bible.

But **we must *mean it from our heart* when we ask Him to forgive our past sins, take control of our life, and be our Lord and Savior now and going forward into eternity** - or our words are meaningless and we're wasting our time saying a prayer of salvation just to say the words to get our *"Get Out of Jail Free"* card, so to speak.

As the Bible says about Salvation (i.e. the way to spend a perfect *eternity* with God), in **Romans 10:9-10, "If you *declare with your mouth,* 'Jesus is Lord,' and *believe in your heart* that God raised him from the dead, you will be *saved.* For it is with your *heart* that you *believe* and are *justified,* and it is with your *mouth* that you *profess your faith* and are *saved."***

So, to be truly *saved* (Christian slang for making Jesus your *Lord and Savior* and spending a perfect eternity with Him)... **we can't simply only *believe* he is who he said he was (and is).** The *Bible says* that even Satan (actually *all* of the demons, which are a third of the original angels) believe that. **Rather, we must *surrender* our life to God through Jesus as well as just believing it. THIS IS THE GREATEST THING YOU WILL EVER DO IN YOUR LIFETIME!! Don't Miss it. How to do this is coming up shortly.**

Until you do this act of surrender, you can't possibly understand what it's all about. In fact, it just sounds either *too good to be true,* or like *a bunch of malarkey.* Atheism and evolution preach *against* this in their secular pseudo-pulpits like TV, movies, radio, newspapers, magazines, science classes in school, etc. They tell you that there is no God or plan of salvation, and that Jesus never died on the cross for YOU. Are you still buying these lies? Don't fall for their half-baked baloney.

Instead, give your life to Jesus *today* and *start living for Him.* It may sound boring, complicated, or too controlled, as if it's going to take all of your freedom away... but in reality it's **Extreme FREEDOM like you've *never* felt before!!** But, you won't know that until you commit your life to Him. It's free, He doesn't want any money, just YOU. Don't delay, do it TODAY (a prayer to pray is coming up here shortly, so prepare your heart to meet your true Creator here and now. TODAY!! Basically just be open minded. Yes, it's OK to get excited. It's not weird. *It's what you and I were meant to do from the beginning.* I'm excited for you)!!

When you sincerely do this, you're inviting *Jesus (*the *King of Kings* and *Lord of Lords)* to take control of your life. The Bible says that you *can't* really understand this until *after* you make this confession and profession to God.

"The person *without* the Spirit does not accept the things that come from the Spirit of God but considers them foolishness, and *cannot* understand them because they are discerned *only* through the Spirit." - *1 Corin. 2:14*

When you commit your life to Him, the Bible says that **He gives you the *HOLY SPIRIT* to live "inside" of you.** To many of you reading this, I know this may sound kind of freaky. Chill my friend, it's *AMAZING.* After you give your life to Jesus, things start to get really interesting. In a *good* way, a *God* way. Life changing ways as a matter of fact. No, you're *not* now going to be perfect, but you will be God's child. And no, this is not some kind of cult experience, it's 100% completely from the Bible. Yes, even that dusty one on the shelf that you have been ignoring. It's time to start reading it. I suggest that you start in the gospel of John. Read it every day.

You'll still have some problems, we all do. Don't let them stop you from your new relationship with God.

You'll start to desire to *know* and *live* for Him… more and more as the days pass. **Life will start to come much more ALIVE!!** Really, this is *not* some sales pitch. You'll say to yourself things like, *"Why didn't I do this before?"* You'll want to stop hurtful stuff you were doing to yourself and others, and let Him live in you and through you.

You'll start telling others about the *Jesus* you once denied and ran from. **Some will think you're crazy, but some will want what *you* now have… *a relationship with JESUS!!*** It's REAL my friends. Far more real than anything you know or have ever known. **It's time to stop running *from* Him, and start running *to* Him.**

Salvation Overview… Because our rebellion is against the authority of God over us, there is one, and *only one way* to obtain peace with God (or come into a right relationship with Him), and that is through the way that God Himself has provided.

This is that God (the Father) sent His Son, the Lord and Savior Jesus Christ, to obey God's law perfectly so that His perfect human righteousness could be imputed to us (credited to our account). Then His death on the cross (followed by His Resurrection) was the perfect offering to pay the penalty for our *sins* (Romans 5:8) - our sins are imputed to Him (Isaiah 53:6–10). God can thus justly forgive our sins, and our guilty past can be cancelled because **our debt has been paid by Jesus Christ** (1 Peter 2:24).

Peace with God comes when we acknowledge our sin (1 John 1:9), change our mind enough to want to change direction (Acts 3:19), and believe and accept what Christ has done on our behalf (Ephesians 2:8). This is a spiritual transaction in which Christ through the Holy Spirit comes to *live within us* (John 1:12–13). All this is what is involved in believing in the Lord Jesus Christ (John 3:16), and receiving Him (John 1:12; Romans 10:13; Revelation 3:20). *- Creation.com*

"The time has come," he said. "The kingdom of God has come near. Repent and believe the good news!" *- Mark 1:15*

"For there is one God, and there is *one* mediator between God and men, the man Christ Jesus." *- 1 Timothy 2:5*

INTERESTING SIDE NOTE FACT (Yes, many Christians are indeed *intelligent*)... **The *majority* (65%) of NOBEL PRIZE winners in the 20th century publicly identified themselves as *"CHRISTIANS."*** *- wikipedia.org (Christian Nobel laureates list)*

Isn't it funny how atheists mock Christians for having *blind faith* and believing in an *eternal God* who *created the universe,* yet they believe that *nothingness* and *eternal energy* created the universe, which we have shown is *completely impossible* in so many ways. So, *who* really has the *blind faith?* Only *one* answer is correct.

Whether you choose to believe it or not, if you believe we all evolved from *nothing,* you have been *conned, fooled, scammed, tricked, duped and downright bamboozled.* Yes, I said bamboozled. *:-)* This all started a long, long time ago from *Satan* himself, and he's still pulling people away from God all the time, everyday, in many very creative ways. Evolution is just one of many ways, but it's one of the greatest deceptions of all time. Don't let it keep you from what God has for you... ETERNAL LIFE!!

"The thief *(Satan)* comes only to steal and kill and destroy; I *(Jesus)* have come that you may have life, and have it abundantly."
- *John 10:10*

Or maybe you choose not to believe in an *Intelligent Creator* for *personal* reasons. Maybe you just don't like the idea of not being in charge of your actions, your life.

I hope by now that you *don't* still think you and your loved ones are the product of a myriad of *random accidents*. But rather that you are the W*onderful Creation* of our *Almighty God.*

God's NOT *Dead, He's surely ALIVE!!* Evolution, however, is truly DEAD!!

Note: *Many of the facts in this book are also from excellent resources like Vance Ferrell's EvolutionFacts.com website, Creation.com Ministries, LivingWaters.com, AnswersInGenesis.org, ICR.org, CreationMoments.com, and several others listed earlier in this book. Make sure to check them out.*

JOIN US DAILY ON
FACEBOOK
Please LIKE, COMMENT and SHARE

facebook.com/CharlieWasWrong

If You Still Believe in Evolution After Reading This Book... it's *Not* for Lack of a *Lot* of Evidence *Disproving* it... Rather it's Probably More Likely a - HEART Issue

I know some of you atheists and evolutionists who just read that topic heading are laughing right now. Some will *never* understand or believe in God. They refuse to no matter how strong the evidence is *against* evolution and atheism. I hope and pray this is not you.

If by now you still choose *irrational evolution* over *God creating the world,* it's certainly not due to lack of hundreds of solid facts *disproving* evolution, but rather it's more likely be a *"heart"* issue.

Meaning, you *choose* not to believe in God, consciously or not, maybe because you just don't quite understand all of this, or for the majority of atheists, you just don't want to give up control of your life to Him. You *choose* rather to be in charge of your own decisions, and destiny apart from Him. You're *not* alone, but does that make it right? **The amazing thing, however, is that God *doesn't* force you to believe in Him but gives you *free-will* to do so, but only if *you* choose to. He's not twisting your arm and forcing you.**

When we pursue the study of *epistemology* (the examination of what distinguishes *justified belief* from *opinion*), we discover the absolute truth of *intelligent design* - and the falsehood and opinions of the many *pretentious theories of evolution.*

So, let me ask you... **what's stopping you from asking Jesus to be your Lord and Savior today?** *(I'm starting to sound like a preacher).*

STOP THE NONSENSE... of believing we all *evolved from "Meaningless Nothing."* Monkeys are *NOT* our ancestors. Open your mind, your eyes, and your heart to the loving God who created the world and every living thing in it. The God who created YOU the amazing person that you are. ***Commit your life to Jesus...*** then find a good local church that teaches the Bible, read your Bible and pray daily, and start living the wonderful and eternal life that He designed for you. Your eternity with God can start TODAY! Right NOW!!

"Here I am! I stand at the door and knock. If anyone hears my voice and opens the door, I will come in and eat with that person, and they with me." - *Jesus (Revelation 3:20)*

OK, IT'S "YOUR" TIME...

This part is for those of you who have *"not"* asked *Jesus* to become your personal *Lord and Savior,* yet. Meaning those of you who have not committed your life to Him, as an pre-teen, teen or adult that is... and not simply *infant baptism* (which is not in the Bible anywhere). You know who you are. :-)

The Bible says... **"If you declare with your mouth, "Jesus is Lord," and believe in your heart that God raised him from the dead, you will be *saved.*"** - *Romans 10:9*

You may be saying, *"Saved from WHAT? I'm a good person."* I think you already know. *Saved from an eternity without God.* Away from His perfect eternity as written about in His word, the Bible. The Bible calls it being, "Born Again." ("Jesus said, "You must be born again." - John 3:3). It's a *"spiritual rebirth."* To be blunt, saved from eternity in Hell. That's exactly what the Bible says in Revelation 20 and other places. Being good alone does "not" get you into Heaven.

I love what **Dr Frank Turek** says at his lectures, ***"If Christianity were true, would you become a Christian?"*** He's had people who claim to be *seeking the truth* actually answer, *"No."* If this is *your* answer, too, I ask you to rethink the question and don't let your *worldview* dictate your answer. Rather, why not let *truth* be your guide.

If you're *not* a Jesus follower yet, there's no better time than *right now* to surrender your life to Him. *Completely.* As the Bible says, there's a time to *die-to-self* and *follow God with everything in you.* Jesus died on the cross to cover our sins and make us right with God. He rose 3 days later and lives today with God in Heaven. You can deny that or accept and embrace it with all that is in you, like hundreds of millions (or more) others believers around the world have done over the past 2,000 years. **Pray this prayer from the bottom of your heart** *(think about what you're praying and really mean it or don't pray it at all),* and ask God to forgive your past sins, to change your life to follow Him, making Jesus the Lord and Savior of your life, thus changing your eternal destiny.

THIS is the Biggest and Most Important Decision You'll EVER Make in Your Lifetime!!!

SINCERELY AND HUMBLY PRAY THIS PRAY (and mean it)…

Lord Jesus, I need You. Thank You for dying on the cross for my sins. I open the door of my life and receive You as my Savior and Lord. Thank You for forgiving my sins and giving me eternal life. Take control of the throne of my life. Make me the kind of person You want me to be. In your name I pray. Amen. *- From CRU (formerly Campus Crusade for Christ)*

If you just prayed that prayer and *meant it,* congratulations, you just changed where you'll spend all of eternity. God says that now he'll give you the **Holy Spirit** **to live** *within* **you** because you asked Jesus to take control of your life as your Lord and Savior. He'll help teach you God's ways through prayer, Bible reading, etc. The Bible says that the angels in Heaven are rejoicing right now because of this decision you just made.

NEXT STEPS... tell a family member, the person who gave you this book, and a friend who you know to be a solid born-again believer, that you just gave your life to Jesus. Let us all know at: Facebook.com/CharlieWasWrong.com. Find a good Jesus believing, Bible teaching church in your city and attend every week. Get plugged into a Bible study. Find a good Christian radio station on your car radio, and Pandora or Spotify. It's extremely important that you start praying (simply talking to God) and also reading your Bible as much as you can *every* day. Start to share Jesus with family, friends and co-workers. **Now, go live for HIM with** *everything* **in you!!**

Time to Say Goodbye for Now

Well, I really hope you enjoyed the book and picked up some tidbits you can use when reaching out to fellow *intelligently designed homo sapiens.* I know that some of you have now left evolutionism, atheism, or agnosticism to become creationists, and many of you have even committed your life to Jesus. How wonderful. Best decision ever!! Some of you never will, how sad. This book started out comparing *evolution* to *intelligent design creation by God,* but it's ultimate goal was to lead as many people to Jesus as possible. I hope that if you are not a Jesus follower yet, you will be open to continue to study God's creation, read the Bible, check out churches in your area, and stay open to the idea of committing your life to Him. You are in my prayers.

AMBASSADORS WANTED

<u>YOU</u> ARE NEEDED!! Join the **CHARLIE WAS WRONG MOVEMENT** and *together* let's wipe out Evolution and *Win People to Jesus!* To spend eternity with Him. Who will stand with me and be a... *Bold Warrior for Christ?* **We believe every Christian should have the information in this book, so they can reach the unsaved.**

1)... ORDER COPIES OF THIS BOOK AT: <u>CharlieWasWrong.com</u> and: Share the real truth about creation with *other "Creationists" (family, friends, online influencers, teachers and pastors).* **ALSO** with *every "non-believer"* **you know** such as: *family, friends, co-workers, influencers, students, and don't forget teachers* (The vast majority of skeptics *won't* buy this book, so you'll want to *give* them a copy).

2)... You (or somebody you know) could start a *study group* with this book; at church, your home, a coffee shop. Teach them to *inform* **the believer,** and to *reach* **the lost.** And, to start their own group.

3)... Consider ordering several books *(monthly)* to give-away.

4)... **WE MUST REACH THE <u>YOUTH</u>!! I believe we can get a study group (using this book) in every high school and college in North America within 3 years. Would you pray about buying a copy for some or even every** *Youth Pastor* **in your area (maybe even copies for the youth also), with the idea of them teaching this stuff to their youth group, and** *also* **setting up small groups in local high schools and colleges.** Youth pastors can give a book to every new student who joins the group, as well as other youth pastors they know. We need people like you to step up and help us get this book to *1,000 youth pastors ASAP!! (Here's how realistic this is – if only 50 of you will order 20 books to pass out to 20 youth pastors in your city/area, we'll reach 1,000 youth pastors, who then can reach and* *teach multitudes of youth... NOW, not "someday."* Can we count on YOU to help reach some in *your* city? Please share this book with your *friends* in *other cities,* too, who can also supply youth pastors in their cities as well (Please read the **Special *"Call of Action"* to Youth and Youth Pastors** section. I hope I can count on you to help us get this teaching into every school in North America, and eventually across the world). And, when schools are closed, students can (and should) be reached via the Internet, or in small groups in homes. Just make sure to abide by the law for such things as *social distancing.*

5)... PLEASE SHARE (via: text, facebook messenger, and email) where people can get this book (at <u>CharlieWasWrong.com</u>), <u>AND</u> please join our **Facebook page** (facebook.com/CharlieWasWrong) and please: **Like, Comment, Invite all your friends, and Share Posts on *your* social media pages, regularly.**

6)... Check out **<u>CharlieWasWrong.com</u>** for discounts on large orders for your church or business. Consider ordering 100, 1,000 or even 10,000+ copies to give to local churches so they can give them to their congregations, youth groups and to people at community events.

7)... Join our "**Ambassador's Club"** (Link at <u>CharlieWasWrong.com</u>)

8)... *PLEASE give a **book REVIEW at Amazon.com** (This is <u>very</u> important as it can help us reach many more people. Do it now).*

These ideas can help us get God's creation and salvation message to millions of students (and adults) in the next few years. Together we can change a multitude of lives to live an eternity with our Awesome Creator, GOD. Evolution *does not* and *cannot* offer this incredible and eternal *hope* and *destiny* that us believers have. **Tell your creationist *and* non-creationist friends that they *need to read this book,* that it absolutely crushes all of the theories of evolution, hands down!!**

Fellow Believers... please PRAY that this book reaches millions of people of all ages for Jesus and that many surrender their lives to Him (Or come back to Him). To God be All the Glory.

Time is Ticking Away. Please Don't Delay. Pray and Order Several Copies Today... to Give Away. <u>CharlieWasWrong.com</u>

ALSO... Join our *AMBASSADOR'S CLUB* from the link there.

Special *"Call of Action"* to Youth / Youth Pastors

DOWN WITH DARWINISM... I have a dream of using this book to reach youth in virtually... **EVERY HIGH SCHOOL and COLLEGE in North America** in the next 3 years. Then, globally. I need YOUR help!! Young people are awesome and they are open to discuss evolution and creation. A study shows that **the vast majority of those who are taught** *"biblical creation"* **early on, stick with God in college years and beyond,** HOWEVER, **the vast majority of Christian kids who are taught** *"evolution"* **(and not creation by their parents or at a youth group),** *leave the church* **by their late teens or early twenties.** They are seeking answers. God is their answer. Together, we can (*and* **must**) reach the youth for the Lord.

Start a study group before or after school, or on lunch breaks. The youth can all learn *much* more about Creation Science *(the most important science)* for next to nothing (the nominal cost of this book), *instead of* $100 to $200 in many cases for a school textbook teaching evolutionary science *(the Un-Science)* that is 50 to 100+ years *out of date.* Share it with friends in "other" youth groups, youth pastors, and schools, who in turn can share it with their friends, and in their schools and youth groups and pastors, too. And, when schools are closed, students can (and should) be reached via the Internet, or in small groups in homes. Just make sure to abide by the law for such things as *social distancing.* Together, with God's help, we can spread the *real story* about Creation and our Creator God, and change the twisted world we now live in, to trust Jesus as their Lord and Savior. *Please join the **"Ambassador's Club"** too (link to join found at CharlieWasWrong.com), so we can all keep in touch and share testimonials on what God is doing in lives.

Chapter Two

Once Upon a Time... the theories of evolution started to fade away under the overwhelming *evidence* for God's true creation story, and many more people lived happily ever after.

Evolution has been a long nightmare. Time to wake up, Sunshine.

The End

The Real Story of How Creation Actually Started

GENESIS 1 - In the *beginning*... God created the Heaven and the Earth. 2 And the Earth was without form, and void; and darkness was upon the face of the deep. And the Spirit of God moved upon the face of the waters. **3** And God said, Let there be light: and there was light. **4** And God saw the light, that it was good: and God divided the light from the darkness. **5** And God called the light Day, and the darkness he called Night. And the evening and the morning were the first day.

6 And God said, Let there be a firmament in the midst of the waters, and let it divide the waters from the waters. **7** And God made the firmament, and divided the waters which were under the firmament from the waters which were above the firmament: and it was so. **8** And God called the firmament Heaven. And the evening and the morning were the second day. **9** And God said, Let the waters under the heaven be gathered together unto one place, and let the dry land appear: and it was so. **10** And God called the dry land Earth; and the gathering together of the waters called them Seas: and God saw that it was good. **11** And God said, Let the Earth bring forth grass, the herb yielding seed, and the fruit tree yielding fruit after his kind, whose seed is in itself, upon the Earth: and it was so. **12** And the Earth brought forth grass, and herb yielding seed after his kind, and the tree yielding fruit, whose seed was in itself, after his kind: and God saw that it was good. **13** And the evening and the morning were the third day. **14** And God said, Let there be lights in the firmament of the heaven to divide the day from the night; and let them be for signs, and for seasons, and for days, and years: **15** And let them be for lights in the firmament of the heaven to give light upon the Earth: and it was so. **16** And God made two great lights; the greater light to rule the day, and the lesser light to rule the night: he made the stars also. **17** And God set them in the firmament of the heaven to give light upon the Earth, **18** And to rule over the day and over the night, and to divide the light from the darkness: and God saw that it was good. **19** And the evening and the morning were the fourth day. **20** And God said, Let the waters bring forth abundantly the moving creature that hath life, and fowl that may fly above the Earth in the open firmament of heaven. **21** And God created great whales, and every living creature that moveth, which the waters brought forth abundantly, after their kind, and every winged fowl after his kind: and God saw that it was good. **22** And God blessed them, saying, Be fruitful, and multiply, and fill the waters in the seas, and let fowl multiply in the Earth. **23** And the evening and the morning were the fifth day. **24** And God said, Let the Earth bring forth the living creature after his kind, cattle, and creeping thing, and beast of the Earth after his kind: and it was so. **25**

And God made the beast of the Earth after his kind, and cattle after their kind, and every thing that creepeth upon the Earth after his kind: and God saw that it was good. **26** And God said, Let us make man in our image, after our likeness: and let them have dominion over the fish of the sea, and over the fowl of the air, and over the cattle, and over all the Earth, and over every creeping thing that creepeth upon the Earth. **27** So God created man in his own image, in the image of God created he him; male and female created he them. **28** And God blessed them, and God said unto them, Be fruitful, and multiply, and replenish the Earth, and subdue it: and have dominion over the fish of the sea, and over the fowl of the air, and over every living thing that moveth upon the Earth. **29** And God said, Behold, I have given you every herb bearing seed, which is upon the face of all the Earth, and every tree, in which is the fruit of a tree yielding seed; to you it shall be for meat. **30** And to every beast of the Earth, and to every fowl of the air, and to every thing that creepeth upon the Earth, wherein there is life, I have given every green herb for meat: and it was so. **31** And God saw every thing that he had made, and, behold, it was very good. And the evening and the morning were the sixth day.

Chapter 2

1 Thus the heavens and the Earth were finished, and all the host of them. 2 And on the seventh day God ended his work; and he rested on the seventh day from all his work which he had made. **3** And God blessed the seventh day, and sanctified it: because that in it he had rested from all his work which God created and made. **4** These are the generations of the heavens and of the Earth when they were created, in the day that the Lord God made the Earth and the heavens. **5** And every plant of the field before it was in the Earth, and every herb of the field before it grew: for the Lord God had not caused it to rain upon the Earth, and there was not a man to till the ground. **6** But there went up a mist from the Earth, and watered the whole face of the ground. **7 And the Lord God formed man of the dust of the ground, and breathed into his nostrils the breath of life; and man became a living soul.** - *King James Version (Public Domain)*

For God so loved the world
that he gave his one and only
Son, that whoever believes
in him shall not perish
but have eternal life.

- John 3:16

Nick Hetcher... lives in Wisconsin with his wife Lynn. They have 4 grown kids (one is Nick's daughter), and 15 grandkids and counting. Hetcher has worked personally with football legend Reggie White, TV infomercial star Don LaPre, and NBC's Last Comic Standing winners Alonzo Bodden and John Heffron. He's also been a voice-over actor for many years. He's been fascinated by God's creation for many years and felt now is the time for him to write this book to debunk the theory of evolution, and share the truth about God's *intelligent design*.

IMPORTANT STUFF

JOIN US ON FACEBOOK
@CharlieWasWrong

Don't forget to check out our
FUN CREATION
T-SHIRTS and STUFF
CharlieWasWrong.com

I Need <u>Your</u> Help with a **"Review"**

I really hope you will take the time to go to <u>Amazon.com</u> and give my book a short review, please. This can be a great way to help us get more books out to more people. Please do it now.

ALSO, I need some of you to send a photo of you holding this book (with a short review attached). And/or a short 10 to 30 second video holding the book. Send them to nickhetcher@gmail.com. Thank you friends.

www.ingramcontent.com/pod-product-compliance
Lightning Source LLC
Chambersburg PA
CBHW021350210526
45463CB00001B/51